John Slater, Thomas Roger Smith

Architecture

Classic and Early Christian

John Slater, Thomas Roger Smith

Architecture
Classic and Early Christian

ISBN/EAN: 9783337179281

Printed in Europe, USA, Canada, Australia, Japan

Cover: Foto ©berggeist007 / pixelio.de

More available books at **www.hansebooks.com**

ILLUSTRATED HANDBOOKS OF ART HISTORY
OF ALL AGES.

ARCHITECTURE

CLASSIC AND EARLY CHRISTIAN

BY PROFESSOR T. ROGER SMITH, F.R.I.B.A.

AND

JOHN SLATER, B.A., F.R.I.B.A.

ILLUSTRATED HAND-BOOKS OF ART HISTORY OF ALL AGES AND COUNTRIES.

EDITED BY

E. J. POYNTER, R.A., Professor ROGER SMITH, F.R.I.B.A., and others.

EACH IN CROWN 8VO, CLOTH EXTRA, PER VOLUME, 5s.

ARCHITECTURE : CLASSIC AND EARLY CHRISTIAN. By Professor T. ROGER SMITH and JOHN SLATER, B.A. Comprising the Egyptian, Assyrian, Greek, Roman, Byzantine, and Early Christian. Illustrated with 200 Engravings, including the Parthenon, and Erechtheum at Athens ; Colosseum, Baths of Diocletian at Rome ; Saint Sophia at Constantinople ; the Sakhra Mosque at Jerusalem, &c.

ARCHITECTURE : GOTHIC AND RENAISSANCE. By Professor T. ROGER SMITH and EDWARD J. POYNTER, R.A. Showing the Progress of Gothic Architecture in England, France, Germany, Italy, and Spain, and of Renaissance Architecture in the same Countries. Illustrated with 100 Engravings, including many of the principal Cathedrals, Churches, Palaces, and Domestic Buildings on the Continent.

SCULPTURE : EGYPTIAN, ASSYRIAN, GREEK, AND RO-MAN. By GEORGE REDFORD, F.R.C.S. With 160 Illustrations of the most celebrated Statues and Bas-Reliefs of Greece and Rome, a Map of Ancient Greece, and a Chronological List of Ancient Sculptors and their Works.

SCULPTURE : GOTHIC, RENAISSANCE AND MODERN. By LEADER SCOTT. Illustrated with numerous Engravings of Works by Ghiberti, Donatello, Della Robbia, Michelangelo, Cellini, and other celebrated Sculptors of the Renaissance. And with Examples of Canova, Thorwaldsen, Flaxman, Chantrey, Gibson, and other Sculptors of the 18th and 19th centuries.

PAINTING : CLASSIC AND ITALIAN. By EDWARD J. POYNTER, R.A., and PERCY R. HEAD, B.A. Including Painting in Egypt, Greece, Rome, and Pompeii ; the Renaissance in Italy : Schools of Florence, Siena, Rome, Padua, Venice, Perugia, Ferrara, Parma, Naples, and Bologna. Illustrated with 50 Engravings of many of the finest Pictures of Italy.

PAINTING : SPANISH AND FRENCH. By GERARD SMITH, Exeter Coll., Oxon. Including the Lives of Ribera, Zurbaran, Velazquez, and Murillo ; Poussin, Claude Lorrain, Le Sueur, Watteau, Chardin, Greuze, David, and Prud'hon ; Ingres, Vernet, Delaroche, and Delacroix ; Corot, Diaz, Rousseau, and Millet ; Courbet, Regnault, Troyon, and many other celebrated artists. With about 80 Illustrations.

PAINTING : GERMAN, FLEMISH, AND DUTCH. By H. J. WILMOT BUXTON, M.A., and EDWARD J. POYNTER, R.A. Including an account of the Works of Albrecht Dürer, Cranach, and Holbein ; Van Eyck, Van der Weyden, and Memline ; Rubens, Snyders, and Van Dyck ; Rembrandt, Hals, and Jan Steen ; Wynants, Ruisdael, and Hobbema ; Cuyp, Potter, and Berchem ; Bakhuisen, Van de Velde, Van Huysum, and other celebrated Painters. Illustrated with 100 Engravings.

PAINTING : ENGLISH AND AMERICAN. By H. J. WILMOT BUXTON, M.A., and S. R. KOEHLER. Including an Account of the Earliest Paintings known in England ; the works of Holbein, Antonio Moro, Lucas de Heere, Zuccaro, and Marc Garrard ; the Hilliards and Olivers ; Van Dyck, Lely, and Kneller ; Hogarth, Reynolds, and Gainsborough ; West, Romney, and Lawrence ; Constable, Turner, and Wilkie ; Maclise, Mulready, and Landseer, and other celebrated Painters. With a Chapter on Painting in America. With 80 Illustrations.

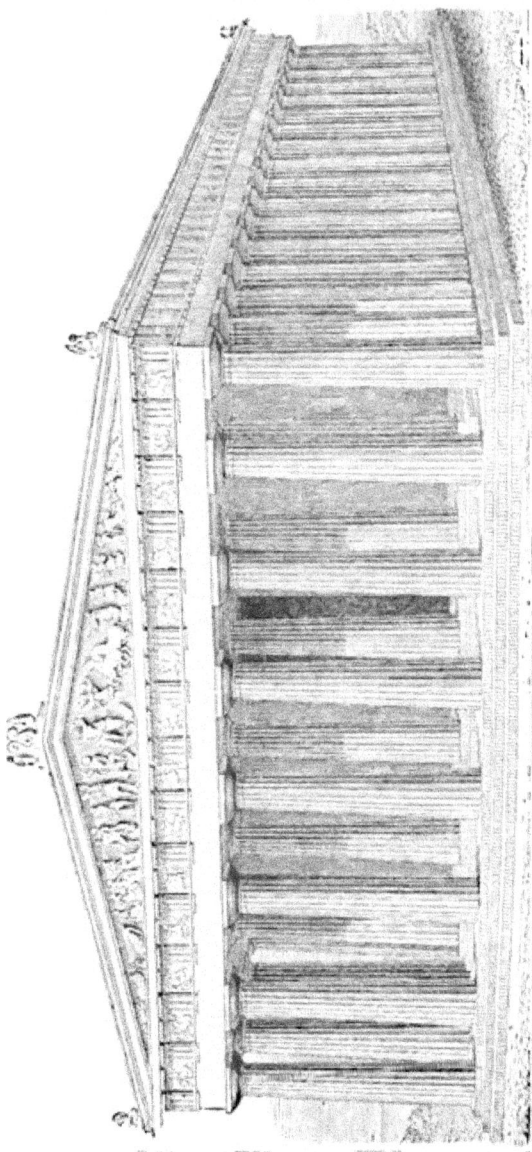

THE PARTHENON AT ATHENS, AS IT WAS IN THE TIME OF PERICLES, *circa* B.C. 438.

ARCHITECTURE

CLASSIC AND EARLY CHRISTIAN

BY T. ROGER SMITH, F.R.I.B.A.
Professor of Architecture, University Coll., London

AND

JOHN SLATER, B.A., F.R.I.B.A.

ATRIUM OF A ROMAN MANSION.

LONDON
SAMPSON LOW, MARSTON, SEARLE, & RIVINGTON, LTD.
St. Dunstan's House
FETTER LANE, FLEET STREET, E.C.
1888

Richard Clay & Sons, Limited, London & Bungay.

PREFACE.

This handbook is intended to give such an outline of the Architecture of the Ancient World, and of that of Christendom down to the period of the Crusades, as, without attempting to supply the minute information required by the professional student, may give a general idea of the works of the great building nations of Antiquity and the Early Christian times. Its chief object has been to place information on the subject within the reach of those persons of literary or artistic education who desire to become in some degree acquainted with Architecture. All technicalities which could be dispensed with have been accordingly excluded; and when it has been unavoidable that a technical word or phrase should occur, an explanation has been added either in the text or in the glossary; but as this volume and the companion one on Gothic and Renaissance Architecture are, in effect, two divisions of the same work, it has not been thought necessary to repeat in the glossary given with this part the words explained in that prefixed to the other.

In treating so very wide a field, it has been felt that the chief prominence should be given to that great sequence of architectural styles which form the links of a chain connecting the architecture of modern Europe with the earliest specimens of the art. Egypt, Assyria, and Persia combined to furnish the foundation upon which the splendid architecture of the Greeks was based.

Roman architecture was founded on Greek models with the
addition of Etruscan construction, and was for a time
universally prevalent. The break-up of the Roman
Empire was followed by the appearance of the Basilican,
the Byzantine, and the Romanesque phases of Christian
art; and, later on, by the Saracenic. These are the styles
on which all mediæval and modern European architecture
has been based, and these accordingly have furnished the
subjects to which the reader's attention is chiefly directed.
Such styles as those of India, China and Japan, which lie
quite outside this series, are noticed much more briefly; and
some matters—such, for example, as prehistoric architecture
—which in a larger treatise it would have been desirable
to include, have been entirely left out for want of room.

In treating each style the object has not been to mention every phase of its development, still less every building, but rather to describe the more prominent buildings
with some approach to completeness. It is true that
much is left unnoticed, for which the student who wishes
to pursue the subject further will have to refer to the
writings specially devoted to the period or country. But
it has been possible to describe a considerable number of
typical examples, and to do so in such a manner as, it is
hoped, may make some impression on the reader's mind.
Had notices of a much greater number of buildings been
compressed into the same space, each must have been so
condensed that the volume, though useful as a catalogue
for reference, would have, in all probability, become
uninteresting, and consequently unserviceable to the class
of readers for whom it is intended.

As far as possible mere matters of opinion have been
excluded from this handbook. A few of the topics which
it has been necessary to approach are subjects on which

high authorities still more or less disagree, and it has been impossible to avoid these in every instance; but, as far as practicable, controverted points have been left untouched. Controversy is unsuited to the province of such a manual as this, in which it is quite sufficient for the authors to deal with the ascertained facts of the history which they have to unfold.

It is not proposed here to refer to the authorities for the various statements made in these pages, but to this rule it is impossible to avoid making one exception. The writers feel bound to acknowledge how much they, in common with all students of the art, are indebted to the patient research, the profound learning, and the admirable skill in marshalling facts displayed by Mr. Fergusson in his various writings. Had it been possible to devote a larger space to Eastern architecture, Pagan and Mohammedan, the indebtedness to him, in a field where he stands all but alone, must of necessity have been still greater.

The earlier chapters of this volume were chiefly written by Mr. Slater, who very kindly consented to assist in the preparation of it; but I am of course, as editor, jointly responsible with him for the contents. The Introduction, Chapters V. to VII., and from Chapter X. to the end, have been written by myself: and if our work shall in any degree assist the reader to understand, and stimulate him to admire, the architecture of the far-off past; above all, if it enables him to appreciate our vast indebtedness to Greek art, and in a lesser degree to the art of other nations who have occupied the stage of the world, the aim which the writers have kept in view will not have been missed.

T. ROGER SMITH.

University College, London.
May, 1882.

FRIZZE FROM CHURCH AT DENKENDORF.

CONTENTS.

LIST OF ILLUSTRATIONS.

b

ROCK-CUT TOMB AT MYRA, IN LYCIA.
Imitation of Timber Construction in Stone.

GLOSSARY.

ABACUS, a square tablet which crowns the capital of the column.

ACANTHUS, a plant the foliage of which was imitated in the ornament of the Corinthian capital.

AGORA, the place of general assembly in a Greek city.

ALÆ (*Lat.* wings), recesses opening out of the atrium of a Roman house.

ALHAMBRA, the palatial fortress of Granada (from *al hamra*—the red).

AMBO, a fitting of early Christian churches, very similar to a pulpit.

AMPHITHEATRE, a Roman place of public entertainment in which combats of gladiators, &c. were exhibited.

ANTÆ, narrow piers used in connection with columns in Greek architecture, for the same purpose as pilasters in Roman.

ARABESQUE, a style of very light ornamental decoration.

ARCHAIC, primitive; so ancient as to be rude, or at least extremely simple.

ARCHIVOLT, the series of mouldings which is carried round an arch.

ARENA, the space in the centre of an amphitheatre where the combats, &c. took place.

ARRIS, a sharp edge.

ASTRAGAL, a small round moulding.

ATRIUM, the main quadrangle in a Roman dwelling-house; also the enclosed court in front of an early Christian basilican church.

BAPTISTERY, a building, or addition to a building, erected for the purposes of celebrating the rite of Christian baptism.

BASEMENT, the lowest story of a building, applied also to the lowest part of an architectural design.

BAS-RELIEF, a piece of sculpture in low relief.

BIRD'S-BEAK, a moulding in Greek architecture, used in the capitals of Antæ.

BYZANTINE, the style of Christian architecture which had its origin at Byzantium (Constantinople).

CARCERES, in the ancient racecourses, goals and starting-points.
CARTOUCHE, in Egyptian buildings, a hieroglyphic signifying the name of a king or other important person.
CARYATIDÆ, human figures made to carry an entablature, in lieu of columns in some Classic buildings.
CAVÆDIAM, another name for the atrium of a Roman house.
CAVEA, the part of an ancient theatre occupied by the audience.
CAVETTO, in Classic architecture, a hollow moulding.
CELLA, the principal, often the only, apartment of a Greek or Roman temple.
CHAITYA, an Indian temple, or hall of assembly.
CIRCUS, a Roman racecourse.
CLOACA, a sewer or drain.
COLUMBARIUM, literally a pigeon-house—a Roman sepulchre built in many compartments.
COLUMNAR, made with columns.
COMPLUVIUM, the open space or the middle of the roof of a Roman atrium.
CORONA, in the cornices of Greek and Roman architecture, the plain unmoulded feature which is supported by the lower part of the cornice, and on which the crowning mouldings rest.
CORNICE, the horizontal series of mouldings crowning the top of a building or the walls of a room.
CUNEIFORM, of letters in Assyrian inscriptions, wedge-shaped.
CYCLOPEAN, applied to masonry constructed of vast stones, usually not hewn or squared.
CYMA (recta, or reversa), a moulding, in Classic architecture, of an outline partly convex and partly concave.

DAGOBA, an Indian tomb of conical shape.
DENTIL-BAND, in Classic architecture, a series of small blocks resembling square-shaped teeth.
DOMUS (Lat.), a house, applied usually to a detached residence.
DWARF-WALL, a very low wall.

ECHINUS, in Greek Doric architecture, the principal moulding of the capital placed immediately under the abacus.
ENTABLATURE, the superstructure—comprising architrave, frieze and cornice—above the columns in Classic architecture.

ENTASIS, in the shaft of a column, a curved outline.

EPHEBEUM, the large hall in Roman baths in which youths practised gymnastic exercises.

FACIA, in Classic architecture, a narrow flat band or face.

FAUCES, the passage from the atrium to the peristyle in a Roman house.

FLUTES, the small channels which run from top to bottom of the shaft of most columns in Classic architecture.

FORUM, the place of general assembly in a Roman city, as the Agora was in a Greek.

FRESCO, painting executed upon a plastered wall while the plaster is still wet.

FRET, an ornament made up of squares and L-shaped lines, in use in Greek architecture.

GARTH, the central space round which a cloister is carried.

GIRDER, a beam.

GROUTED, said of masonry or brickwork, treated with liquid mortar to fill up all crevices and interstices.

GUTTÆ, small pendent features in Greek and Roman Doric cornices, resembling rows of wooden pegs.

HEXASTYLE, of six columns.

HONEYSUCKLE ORNAMENT, a decoration constantly introduced into Assyrian and Greek architecture, founded upon the flower of the honeysuckle.

HORSESHOE ARCH, an arch more than a semicircle, and so wider above than at its springing.

HYPOSTYLE, literally "under columns," but used to mean filled by columns.

IMPLUVIUM, the space into which the rain fell in the centre of the atrium of a Roman house.

INSULA, a block of building surrounded on all sides by streets, literally an island.

INTERCOLUMNIATION, the space between two columns.

KEYED, secured closely by interlocking.

KIBLA, the most sacred part of a Mohammedan mosque.

LÂTS, in Indian architecture, Buddhist inscribed pillars.

MAMMISI, small Egyptian temples.

MASTABA, the most usual form of Egyptian tomb.

MAUSOLEUM, a magnificent sepulchral monument or tomb. From the tomb erected to Mausolus, by his wife Artemisia, at Halicarnassus, 379 B.C.

METOPES, literally faces, the square spaces between triglyphs in Doric architecture ; occasionally applied to the sculptures fitted into these spaces.

MINARET, a slender lofty tower, a usual appendage of a Mohammedan mosque.

MONOLITH, of one stone.

MORTISE, a hollow in a stone or timber to receive a corresponding projection.

MOSQUE, a Mohammedan place of worship.

MUTULE, a feature in a Classic Doric cornice, somewhat resembling the end of a timber beam.

NARTHEX, in an early Christian church, the space next the entrance.

OBELISK, a tapering stone pillar, a feature of Egyptian architecture.

OPUS ALEXANDRINUM, the mosaic work used for floors in Byzantine and Romanesque churches.

OVOLO, a moulding, the profile of which resembles the outline of an egg, used in Classic architecture.

PENDENTIVE, a feature in Byzantine and other domed buildings, employed to enable a circular dome to stand over a square space.

PERISTYLAR, or PERIPTERAL, with columns all round.

PERISTYLIUM, or PERISTYLE, in a Roman house, the inner courtyard ; also any space or enclosure with columns all round it.

PISCINA, a small basin usually executed in stone and placed within a sculptured niche, fixed at the side of an altar in a church, with a channel to convey away the water poured into it.

POLYCHROMY, the use of decorative colours.

PRECINCTS, the space round a church or religious house, usually enclosed with a wall.

PRESBYTERY, the eastern part of a church, the chancel.

PROFILE (of a moulding), the outline which it would present if cut across at right angles to its length.

PRONAOS, the front portion or vestibule to a temple.

PROPYLÆA, in Greek architecture, a grand portal or state entrance.

PROTHYRUM, in a Roman house, the porch or entrance.

PSEUDO-PERIPTERAL, resembling, but not really being peristylar.

PYLON, or PRO-PYLON, the portal or front of an Egyptian temple.

QUADRIGA, a four-horse chariot.

ROMANESQUE, the style of Christian architecture which was founded on Roman work.

ROTUNDA, a building circular in plan.

SACRISTY, the part of a church where the treasures belonging to the church are preserved.

SHINTO TEMPLES, temples (in Japan) devoted to the Shinto religion.

SPAN, the space over which an arch or a roof extends.

SPINA, the central wall of a Roman racecourse.

STILTED, raised, usually applied to an arch when its centre is above the top of the jambs from which it springs.

STRUTS, props.

STUPA, in Indian architecture, a mound or tope.

STYLOBATE, a series of steps, usually those leading up to a Classic temple.

TAAS, a pagoda.

TABLINUM, in a Roman house, the room between the atrium and the peristyle.

TALAR, in Assyrian architecture, an open upper story.

TENONED, fastened with a projection or tenon.

TESSELATED, made of small squares of material, applied to coarse mosaic work.

TETRASTYLE, with four columns.

THERMÆ, the great bathing establishments of the Romans.

TOPES, in Indian architecture, artificial mounds.

TRABEATED, constructed with a beam or beams, a term usually employed in contrast to arches.

TRICLINIUM, in a Roman house, the dining-room.

TRIGLYPH, the channelled feature in the frieze of the Doric order.

TUMULI, mounds, usually sepulchral.

TYPHONIA, small Egyptian temples.

VELARIUM, a great awning.

VESTIBULE, the outer hall or ante-room.

VOLUTES, in Classic architecture, the curled ornaments of the Ionic capital.

VOUSSOIRS, the wedge-shaped stones of which arches are made.

N.B. For the explanation of other technical words found in this volume, consult the Glossary given with the companion volume on Gothic and Renaissance Architecture.

THE TEMPLE OF VESTA AT TIVOLI.

ANCIENT ARCHITECTURE.

CHAPTER I.

INTRODUCTION.

ARCHITECTURE may be described as building at its best, and when we talk of the architecture of any city or country we mean its best, noblest, or most beautiful buildings; and we imply by the use of the word that these buildings possess merits which entitle them to rank as works of art.

The architecture of the civilised world can be best understood by considering the great buildings of each important nation separately. The features, ornaments, and even forms of ancient buildings differed just as the speech, or at any rate the literature, differed. Each nation wrote in a different language, though the books may have been devoted to the same aims; and precisely in the same way each nation built in a style of its own, even if the buildings may have been similar in the purposes they had to serve. The division of the subject into the architecture of Egypt, Greece, Rome, &c., is therefore the most natural one to follow.

But certain broad groups, rising out of peculiarities of a physical nature, either in the buildings themselves or in the conditions under which they were erected, can hardly fail to be suggested by a general view of the subject. Such, for example, is the fourfold division to which the reader's attention will now be directed.

All buildings, it will be found, can be classed under one or other of four great divisions, each distinguished by a distinct mode of building, and each also occupying a distinct place in history. The first series embraces the buildings of the Egyptians, the Persians, and the Greeks, and was brought to a pitch of the highest perfection in Greece during the age of Pericles. All the buildings erected in these countries during the many centuries which elapsed from the earliest Egyptian to the latest Greek works, however they may have differed in other respects, agree in this—that the openings, be they doors, or be they spaces between columns, were spanned by beams of wood or lintels of stone (Fig. 1). Hence this architecture is called architecture of the beam, or, in more formal language, trabeated architecture. This mode of covering spaces required that in buildings of solid masonry, where stone or marble lintels were employed, the supports should not be very far apart, and this circumstance led to the frequent use of rows of columns. The architecture of this period is accordingly sometimes called columnar, but it has no exclusive claim to the epithet; the column survived long after the exclusive use of the beam had been superseded, and the term

FIG. 1.—OPENING SPANNED BY A LINTEL. ARCH OF THE GOLDSMITHS, ROME.

columnar must accordingly be shared with buildings forming part of the succeeding series.

The second great group of buildings is that in which the semicircular arch is introduced into construction, and

used either together with the beam, or, as mostly happened,
instead of the beam, to span the openings (Fig. 2). This
use of the arch began with the Assyrians, and it reap-
peared in the works of the early Etruscans. The round-
arched series of styles embraces the buildings of the
Romans from their earliest beginnings to their decay; it
also includes the two great schools of Christian architec-

FIG. 2.—OPENING SPANNED BY A SEMICIRCULAR ARCH. ROMAN TRIUMPHAL ARCH
AT POLA.

ture which were founded by the Western and the Eastern
Church respectively,—namely, the Romanesque, which,
originating in Rome, extended itself through Western
Europe, and lasted till the time of the Crusades, and the
Byzantine, which spread from Constantinople over all
the countries in which the Eastern (or Greek) Church
flourished, and which continues to our own day.

The third group of buildings is that in which the pointed arch is employed instead of the semicircular arch to span the openings (Fig. 3). It began with the rise of

Fig. 3.—Openings spanned by Pointed Arches. Interior of St. Front, Perigueux, France.

Mohammedan architecture in the East, and embraces all the buildings of Western Europe, from the time of the First Crusade to the revival of art in the fifteenth century.

This great series of buildings constitutes what is known
as Pointed, or, more commonly, as Gothic architecture.

The fourth group consists of the buildings erected
during or since the Renaissance (*i. e.* revival) period, and
is marked by a return to the styles of past ages or distant
countries for the architectural features and ornaments of
buildings; and by that luxury, complexity, and ostenta-
tion which, with other qualities, are well comprehended
under the epithet Modern. This group of buildings
forms what is known as Renaissance architecture, and
extends from the epoch of the revival of letters in the
fifteenth century, to the present day.

The first two of these styles—namely, the architecture
of the beam, and that of the round arch—are treated of in
this little volume. They occupy those remote times of
pagan civilisation which may be conveniently included
under the broad term Ancient; and the better known
work of the Greeks and Romans—the classic nations—
and they extend over the time of the establishment of
Christianity down to the close of that dreary period
not incorrectly termed the Dark ages. Ancient, Classic,
and early Christian architecture is accordingly an appropriate title for the main subjects of this volume, though,
for the sake of convenience, some notices of Oriental
architecture have been added. Gothic and Renaissance
architecture form the subjects of the companion volume.

It may excite surprise that what appears to be so small
a difference as that which exists between a beam, a round
arch, or a pointed arch, should be employed in order to
distinguish three of the four great divisions. But in reality
this is no pedantic or arbitrary grouping. The mode in
which spaces or openings are covered lies at the root of
most of the essential differences between styles of archi-

tecture, and the distinction thus drawn is one of a real, not of a fanciful nature.

Every building when reduced to its elements, as will be done in both these volumes, may be considered as made up of its (1) floor or plan, (2) walls, (3) roof, (4) openings, (5) columns, and (6) ornaments, and as marked by its distinctive (7) character, and the student must be prepared to find that the openings are by no means the least important of these elements. In fact, the moment the method of covering openings was changed, it would be easy to show, did space permit, that all the other elements, except the ornaments, were directly affected by the change, and the ornaments indirectly ; and we thus find such a correspondence between this index feature and the entire structure as renders this primary division a scientific though a very broad one. The contrast between the trabeated style and the arched style may be well understood by comparing the illustration of the Parthenon which forms our frontispiece, or that of the great temple of Zeus at Olympia (Fig. 4), with the exterior of the Colosseum at Rome (Fig. 5), introduced here for the purposes of this comparison.

A division of buildings into such great series as these cannot, however, supersede the more obvious historical and geographical divisions. The architecture of every ancient country was partly the growth of the soil, i.e. adapted to the climate of the country, and the materials found there, and partly the outcome of the national character of its inhabitants, and of such influences as race, colonisation, commerce, or conquest brought to bear upon them. These influences produced strong distinctions between the work of different peoples, especially before the era of the Roman Empire. Since that

Fig. 1.—Temple of Zeus at Olympia. Restored according to Adler.

period of universal dominion all buildings and styles have been influenced more or less by Roman art. We accordingly find the buildings of the most ancient nations separated from each other by strongly marked lines of demarcation, but those since the era of the Empire showing a considerable resemblance to one another. The circumstance that the remains of those buildings only which received the greatest possible attention from their builders have come down to us from any remote antiquity, has perhaps served to accentuate the differences between different styles, for these foremost buildings were not intended to serve the same purpose in all countries. Nothing but tombs and temples have survived in Egypt. Palaces only have been rescued from the decay of Assyrian and Persian cities; and temples, theatres, and places of public assembly are the chief, almost the only remains of architecture in Greece.

A strong contrast between the buildings of different ancient nations rises also from the differing point of view for which they were designed. Thus, in the tombs and, to a large extent, the temples of the Egyptians, we find structures chiefly planned for internal effect; that is to say, intended to be seen by those admitted to the sacred precincts, but only to a limited extent appealing to the admiration of those outside. The buildings of the Greeks, on the other hand, were chiefly designed to please those who examined them from without; and though no doubt some of them, the theatres especially, were from their very nature planned for interior effect, by far the greatest works which Greek art produced were the exteriors of the temples.

The works of the Romans, and, following them, those of almost all Western Christian nations, were designed

FIG. 5.—PART OF THE EXTERIOR OF THE COLOSSEUM, ROME. (NOW IN RUINS.)

to unite external and internal effect; but in many cases external was evidently most sought after, and, in the North of Europe, many expedients—such, for example, as towers, high-pitched roofs, and steeples—were introduced into architecture with the express intention of increasing external effect. On the other hand, the Eastern styles, both Mohammedan and Christian, especially when practised in sunny climates, show in many cases a comparative disregard of external effect, and that their architects lavished most of their resources on the interiors of their buildings.

Passing allusions have been made to the influence of climate on architecture; and the student whose attention has been once called to this subject will find many interesting traces of this influence in the designs of buildings erected in various countries. Where the power of the sun is great, flat terraced roofs, which help to keep buildings cool, and thick walls are desirable. Sufficient light is admitted by small windows far apart. Overhanging eaves, or horizontal cornices, are in such a climate the most effective mode of obtaining architectural effect, and accordingly in the styles of all Southern peoples these peculiarities appear. The architecture of Egypt, for example, exhibited them markedly. Where the sun is still powerful, but not so extreme, the terraced roof is generally replaced by a sloping roof, steep enough to throw off water, and larger openings are made for light and air; but the horizontal cornice still remains the most appropriate means of gaining effects of light and shade. This description will apply to the architecture of Italy and Greece. When, however, we pass to Northern countries, where snow has to be encountered, where light is precious, and where the sun is low in the heavens for the

greater part of the day, a complete change takes place.
Roofs become much steeper, so as to throw off snow.
The horizontal cornice is to a large extent disused, but
the buttress, the turret, and other vertical features, from
which a level sun will cast shadows, begin to appear;

FIG. 6.—TIMBER ARCHITECTURE. CHURCH AT BORGUND.

and windows are made numerous and spacious. This
description applies to Gothic architecture generally—in
other words, to the style which rose in Northern Europe.

The influence of materials on architecture is also worth
notice. Where granite, which is worked with difficulty,

is the material obtainable, architecture has invariably
been severe and simple; where soft stone is obtainable,
exuberance of ornament makes its appearance, in consequence of the material lending itself readily to the carver's
chisel. Where, on the other hand, marble is abundant
and good, refinement is to be met with, for no other
building material exists in which very delicate mouldings
or very slight or slender projections may be employed
with the certainty that they will be effective. Where
stone is scarce, brick buildings, with many arches, roughly
constructed cornices and pilasters, and other peculiarities
both of structure and ornamentation, make their appearance, as, for example, in Lombardy and North Germany.
Where materials of many colours abound, as is the case, for
example, in the volcanic districts of France, polychromy
is sought as a means of ornamentation. Lastly, where
timber is available, and stone and brick are both scarce,
the result is an architecture of which both the forms and
the ornamentation are entirely dissimilar to those proper
to buildings of stone, marble, or brick, as may be seen by
a glance at our illustration of an early Scandinavian
church built of timber (Fig. 6), which presents forms
appropriate to a timber building as being easily constructed of wood, but which would hardly be suitable to
any other material whatever.

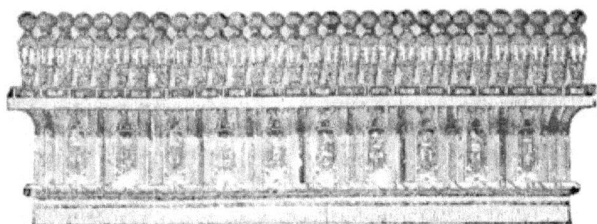

FIG. 7.—EGYPTIAN CORNICE.

CHAPTER II.

EGYPTIAN ARCHITECTURE.

THE origin of Egyptian architecture, like that of Egyptian history, is lost in the mists of antiquity. The remains of all, or almost all, other styles of architecture enable us to trace their rude beginnings, their development, their gradual progress up to a culminating point, and thence their slow but certain decline; but the earliest remains of the constructions of the Egyptians show their skill as builders at the height of its perfection, their architecture highly developed, and their sculpture at its very best, if not indeed at the commencement of its decadence; for some of the statuary of the age of the Pyramids was never surpassed in artistic effect by the work of a later era. It is impossible for us to conceive of such scientific skill as is evidenced in the construction of the great pyramids, or such artistic power as is displayed on the walls of tombs of the same date, or in the statues found in them, as other than the outcome of a vast accumulation of experience, the attainment of which must imply the lapse of very long periods of time since the nation which produced such works emerged from barbarism. It is natural, where so remote an antiquity is in question, that we should feel a great difficulty, if not an impossibility, in fixing exact dates, but the whole tendency of modern exploration and research is rather to push back than to advance the dates of Egyptian chronology, and it is by no means impossible that the dynasties of Manetho, after being derided as apocryphal for centuries, may in the end be accepted as substantially correct. Manetho was an Egyptian priest living in the third century B.C., who wrote a history of his country, which he compiled from the archives of the temples. His work itself is lost, but Josephus quotes extracts from it, and Eusebius and Julius Africanus reproduced his lists, in which the monarchs of Egypt are grouped into thirty-four dynasties. These, however, do not agree with one another, and in many cases it is difficult to reconcile them with the records displayed in the monuments themselves.

The remains with which we are acquainted indicate four distinct periods of great architectural activity in Egyptian history, viz. : (1) the period of the fourth dynasty, when the Great Pyramids were erected (probably 3500 to 3000 B.C.); (2) the period of the twelfth dynasty, to which belong the remains at Beni-Hassan ; (3) the period of the eighteenth and nineteenth dynasties, when Thebes was in its glory, which is attested by the ruins of Luxor and Karnak ; and (4) the Ptolemaic period, of which there are the remains at Denderah, Edfou, and Philæ. The monuments that remain are almost exclusively tombs and temples. The tombs are, generally speaking, all met with on the east or right bank of the Nile: among them must be classed those grandest and oldest monuments of Egyptian skill, the

Pyramids, which appear to have been all designed as
royal burying-places. A large number of pyramids have
been discovered, but those of Gizeh, near Cairo, are the
largest and the best known, and also probably the oldest
which can be authenticated.* The three largest pyramids
are those of Cheops, Cephren, and Mycerinus at Gizeh
(or, as the names are more correctly written, Suphis, Sen-
suphis, and Moscheris or Mencheris). These monarchs
all belonged to the fourth dynasty, and the most probable
date to be assigned to them is about 3000 B.C. The pyramid of Suphis is the largest, and is the one familiarly
known as the Great Pyramid; it has a square base, the
side of which is 760 feet long,† a height of 484 feet, and
an area of 577,600 square feet. In this pyramid the angle
of inclination of the sloping sides to the base is 51° 51′,
but in no two pyramids is this angle the same. There
can be no doubt that these huge monuments were erected
each as the tomb of an individual king, whose efforts were
directed towards making it everlasting, and the greatest
pains were taken to render the access to the burial chamber
extremely hard to discover. This accounts for the vast
disproportion between the lavish amount of material used
for the pyramid and the smallness of the cavity enclosed
in it (Fig. 8).

The material employed was limestone cased with syenite
(granite from Syene), and the internal passages were lined
with granite. The granite of the casing has entirely

* Some Egyptologists incline to the opinion that the pyramid of Saq-
q'ra is the most ancient, while others think it much more recent than
those of Gizeh.

† Strictly speaking, the base is not an exact square, the four sides
measuring, according to the Royal Engineers, north, 760 ft. 7·5 in.; south,
761 ft. 8·5 in.; east, 760 ft. 9·5 in.; and west, 764 ft. 1 in.

disappeared, but that employed as linings is still in its place, and so skilfully worked that it would not be possible to introduce even a sheet of paper between the joints.

The entrance D to this pyramid of Suphis was at a height of 47 ft. 6 in. above the base, and, as was almost invariably

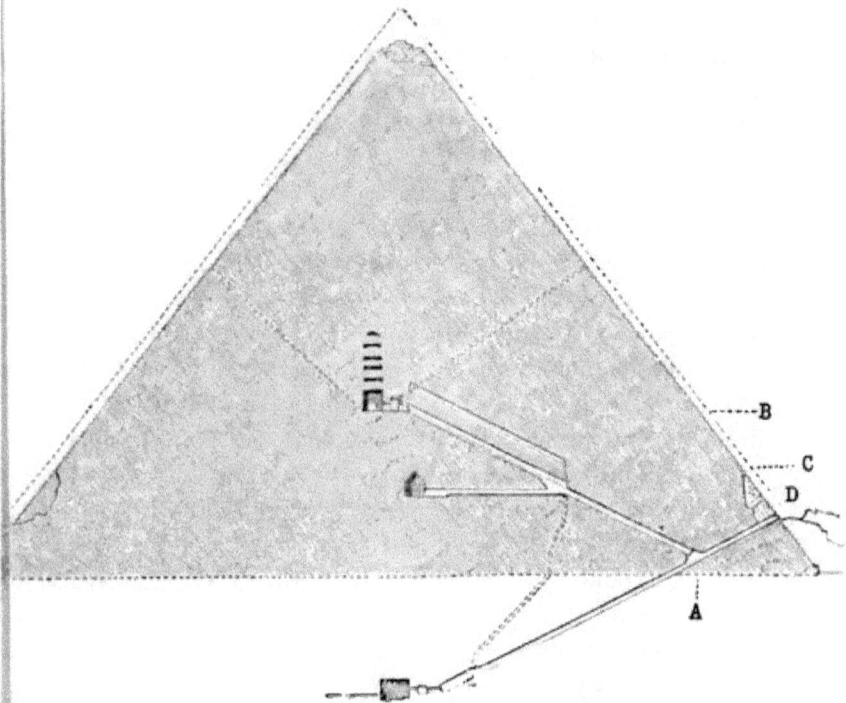

FIG 9.—SECTION ACROSS THE GREAT PYRAMID (OF CHEOPS OR SUPHIS).

the case, on the north face; from the entrance a passage slopes downward at an angle of 26° 27' to a chamber cut in the rock at a depth of about 90 feet below the base of the pyramid. This chamber seems to have been intended as a blind, as it was not the place for the deposition

C

of the corpse. From the point in the above described passage—marked A on our illustration of this pyramid —another gallery starts upwards, till it reaches the point C, from which a horizontal passage leads to another small chamber. This is called the Queen's Chamber, but no reason has been discovered for the name. From this point C the gallery continues upwards till, in the heart of the pyramid, the Royal Chamber, B, is reached. The walls of these chambers and passages are lined with masonry executed in the hardest stone (granite), and with an accuracy of fitting and a truth of surface that can hardly be surpassed. Extreme care seems to have been taken to prevent the great weight overhead from crushing in the galleries and the chamber. The gallery from C upwards is of the form shown in Fig. 9, where each layer of stones projects slightly beyond the one underneath it. Fig. 11 is a section of the chamber itself, and the succession of small chambers shown one above the other was evidently formed for the purpose of distributing the weight of the superincumbent mass. From the point C a narrow well leads almost perpendicularly downwards to a point nearly at the bottom of the first-mentioned gallery; and the purpose to be served by this well was long a subject of debate. The probability is that, after the corpse had been placed in its chamber, the workmen completely blocked up the passage from A to C by allowing large blocks of granite to slide down it, these blocks having been previously prepared and deposited in the larger gallery; the men then let themselves down the well, and by means of the lower gallery made their exit from the pyramid. The entrances to the chamber and to the pyramid itself were formed by huge blocks of stone which exactly fitted into grooves prepared for them with the

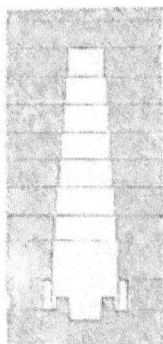

FIG. 9.—ASCENDING GALLERY IN THE GREAT PYRAMID.

FIG. 10.—THE SEPULCHRAL CHAMBER IN THE PYRAMID OF CEPHREN AT GIZEH.

FIG. 11.—THE CONSTRUCTION OF THE KING'S CHAMBER IN THE GREAT PYRAMID.

most beautiful mathematical accuracy. The chief interest attaching to the pyramids lies in their extreme antiquity, and the scientific method of their construction; for their effect upon the spectator is by no means proportionate to their immense mass and the labour bestowed upon them.

In the neighbourhood of the pyramids are found a large number of tombs which are supposed to be those of private persons. Their form is generally that of a *mastaba* or truncated pyramid with sloping walls, and their construction is evidently copied from a fashion of wooden architecture previously existing. The same idea of making an everlasting habitation for the body prevailed as in the case of the pyramids, and stone was therefore the material employed; but the builders seem to have desired to indulge in a decorative style, and as they were totally unable to originate a legitimate stone architecture, we find carved in stone, rounded beams as lintels, grooved posts, and—most curious of all—roofs that are an almost exact copy of the early timber huts when unsquared baulks of timber were laid across side by side to form a covering. Figs. 12 and 13 show this kind of stone-work, which is peculiar to the old dynasties, and seems to have had little influence upon succeeding styles.

A remarkable feature of these early private tombs consists in the paintings with which the walls are decorated, and which vividly portray the ordinary every-day occupations carried on during his lifetime by the person who was destined to be the inmate of the tomb. These paintings are of immense value in enabling us to form an accurate idea of the life of the people at this early age.

It may possibly be open to doubt whether the dignified appellation of architecture should be applied to buildings

FIG. 12.—IMITATION OF TIMBER CONSTRUCTION IN
STONE, FROM A TOMB AT MEMPHIS.

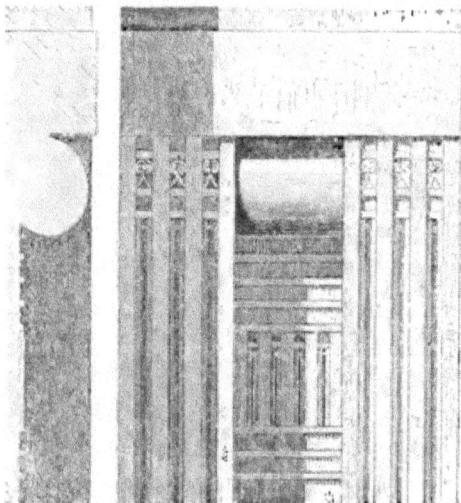

FIG. 13.—IMITATION OF TIMBER CONSTRUCTION IN
STONE, FROM A TOMB AT MEMPHIS.

of the kind we have just been describing; but when we
come to the series of remains of the twelfth dynasty at
Beni-Hassan, in middle Egypt, we meet with the earliest
known examples of that most interesting feature of all
subsequent styles — the column.   Whether the idea of
columnar architecture originated with the necessities of
quarrying—square piers being left at intervals to support
the superincumbent mass of rock as the quarry was gradu-
ally driven in—or whether the earliest stone piers were
imitations of brickwork or of timber posts, we shall probably never be able to determine accurately, though the
former supposition seems the more likely.   We have here
monuments of a date 1400 years anterior to the earliest
known Greek examples, with splendid columns, both
exterior and interior, which no reasonable person can
doubt are the prototypes of the Greek Doric order.
Fig. 14 is a plan with a section, and Fig. 15 an exterior
view, of one of these tombs, which, it will be seen, consisted of a portico, a chamber with its roof supported by
columns, and a small space at the farther end in which
is formed the opening of a sloping passage or well, at the
bottom of which the vault for the reception of the body
was constructed.   The walls of the large chamber are
lavishly decorated with scenes of every-day life, and it
has even been suggested that these places were not erected
originally as tombs, but as dwelling-places, which after
death were appropriated as sepulchres.

The columns are surmounted by a small square slab,
technically called an abacus, and heavy square beams or
architraves span the spaces between the columns, while
the roof between the architraves has a slightly segmental
form.   The tombs of the later period, viz. of the eighteenth
and nineteenth dynasties, are very different from those of

the twelfth dynasty, and present few features of architectural interest, though they are remarkable for their
vast extent and the
variety of form of their
various chambers and
galleries. They con-
sist of a series of
chambers excavated in
the rock, and it ap-
pears certain that the
tomb was commenced
on the accession of each
monarch, and was
driven farther and
farther into the rock
during the continuance
of his reign till his
death, when all work
abruptly ceased. All
the chambers are pro-
fusely decorated with
paintings, but of a kind
very different from
those of the earlier
dynasties. Instead of
depicting scenes of ordi-
nary life, all the paint-
ings refer to the sup-
posed life after death,
and are thus of very
great value as a means
of determining the re-
ligious opinions of the

SECTION.

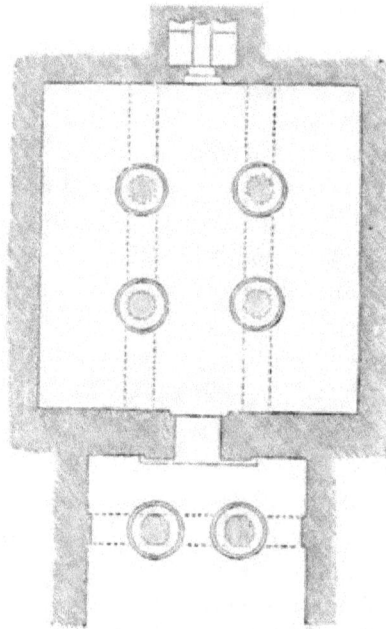

FIG. 14.—PLAN AND SECTION OF THE TOMB
AT BENI-HASSAN.

Egyptians at this time.   One of the most remarkable
of these tombs is that of Manephthah or Sethi 1., at
Bab-el-Molouk, and known as Belzoni's tomb, as it was
discovered by him; from it was taken the alabaster sarcophagus now in the Soane Museum in Lincoln's Inn Fields.
To this relic a new interest is given by the announcement,
while these pages are passing through the press, of the
discovery of the mummy of this very Manephthah, with
thirty-eight other royal mummies, in the neighbourhood
of Thebes.

FIG. 15.—ROCK-CUT FAÇADE OF TOMB AT BENI-HASSAN

Of the Ptolemaic period no tombs, except perhaps a few
at Alexandria, are known to exist.

TEMPLES.

It is very doubtful whether any remains of temples of
the time of the fourth dynasty—i.e. contemporaneous with
the pyramids—exist.   One, constructed on a most extraordinary plan, was supposed to have been discovered about
a quarter of a century ago, and it was described by Pro-

fessor Donaldson at the Royal Institute of British Architects in 1861, but later Egyptologists rather incline to the belief that this was a tomb and not a temple, as in one of the chambers of the interior a number of compartments were discovered one above the other which were apparently intended for the reception of bodies. This singular building is close to the Great Sphinx ; its plan is cruciform, and there are in the interior a number of rectangular piers of granite supporting very simple architraves, but there are no means of determining what kind of roof covered it in. The walls seem to have been faced on the interior with polished slabs of granite or alabaster, but no sculpture or hieroglyphic inscriptions were found on them to explain the purpose of the building. Leaving this building—which is of a type quite unique—out of the question, Egyptian temples can be generally classed under two heads: (1) the large principal temples, and (2) the small subsidiary ones called Typhonia or Mammisi. Both kinds of temple vary little, if at all, in plan from the time of the twelfth dynasty down to the Roman dominion.

The large temples consist almost invariably of an entrance gate flanked on either side by a large mass of masonry, called a pylon, in the shape of a truncated pyramid (Fig. 18). The axis of the ground-plan of these pylons is frequently obliquely inclined to the axis of the plan of the temple itself; and indeed one of the most striking features of Egyptian temples is the lack of regularity and symmetry in their construction. The entrance gives access to a large courtyard, generally ornamented with columns : beyond this, and occasionally approached by steps, is another court, smaller than the first, but much more splendidly adorned with columns and colossi; beyond this again, in the finest examples, occurs what is called the Hypostyle Hall, *i.e.* a hall with two rows of lofty columns down the centre, and at the sides other rows, more or less in number, of lower columns; the object of this arrangement being that the central portion might be lighted by a kind of clerestory above the roof of the side portions. Fig. 17 shows this arrangement. This hypostyle hall stood with its greatest length transverse to the general axis of the temple, so that it was entered from the side. Beyond it were other chambers, all of small size, the

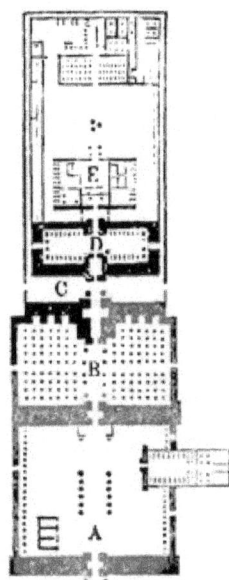

innermost being generally the sanctuary, while the others were probably used as residences by the priests. Homer's hundred-gated Thebes, which was for so long the capital of Egypt, offers at Karnak and Luxor the finest remains of temples; what is left of the former evidently showing that it must have been one of the most magnificent buildings ever erected in any country. Fig. 16 is a plan of the temple of Karnak, which was about 1200 feet long and 348 feet wide. A is the entrance between the two enormous pylons giving access to a large courtyard, in which is a small detached temple, and another larger one breaking into the courtyard obliquely. A gateway between a second pair of pylons admits to B, the grand Hypostyle Hall, 334 feet by 167 feet. Beyond this are additional gateways with pylons, separated by a sort

FIG. 17.—THE HYPOSTYLE HALL AT KARNAK, SHOWING THE CLERESTORY.

FIG. 18.—ENTRANCE TO AN EGYPTIAN TEMPLE, SHOWING THE PYLONS.

of gallery, C, in which were two gigantic obelisks; D, another grand hall, is called the Hall of the Caryatides, and beyond is the Hall of the eighteen columns, through which access is gained to a number of smaller halls grouped round the central chamber E. Beyond this is a large courtyard, in the centre of which stood the original sanctuary, which has disappeared down to its foundations, nothing but some broken shafts of columns remaining. At the extreme east is another hall supported partly by columns and partly by square piers, and a second series of pillared courts and chambers. The pylons and buildings generally decrease in height as we proceed from the entrance eastwards. This is due to the fact that the building grew by successive additions, each one more magnificent than the last, all being added on the side from which the temple was entered, leaving the original sanctuary unchanged and undisturbed.

Besides the buildings shown on the plan there were many other temples to the north, south, and east, entered by pylons and some of them connected together by avenues of sphinxes, obelisks, and colossi, which altogether made up the most wonderful agglomeration of buildings that can be conceived. It must not be imagined that this temple of Karnak, together with the series of connected temples is the result, of one clearly conceived plan; on the contrary, just as has been frequently the case with our own cathedrals and baronial halls, alterations were made here and additions there by successive kings one after the other without much regard to connection or congruity, the only feeling that probably influenced them being that of emulation to excel in size and grandeur the erections of their predecessors, as the largest buildings are almost always of latest date. The original sanctuary, or nucleus of the temple, was built by Usertesen I., the second or third king of the twelfth dynasty. Omenophis, the first king of the Shepherd dynasties, built a temple round the sanctuary, which has disappeared. Thothmes I. built the Hall of the Caryatides and commenced the next Hall of the eighteen columns, which was finished by Thothmes II. Thothmes III. built that portion surrounding the sanctuary, and he also built the courts on the extreme east. The pylon at C was built by Omenophis III., and formed the façade of the temple before the erection of the grand hall. Sethi I. built the Hypostyle Hall, which had probably been originated by Rhamses I., who commenced the pylon west of it. Sethi II. built the small detached temple, and Rhamses III. the intersecting temple. The Bubastites constructed the large front court by building walls round it, and the Ptolemies commenced the huge western pylon. The colonnade in the centre of the court was erected by Tahraka.

Extensive remains of temples exist at Luxor, Edfou (Fig. 19), and Philæ, but it will not be necessary to give a detailed description of them, as, if smaller in size, they are very similar in arrangement to those already described. It should be noticed that all these large temples have the mastaba form, *i.e.* the outer walls are not perpendicular on the outside, but slope inwards as they rise, thus giving the buildings an air of great solidity.

The Mammisi exhibit quite a different form of temple from those previously described, and are generally found in close proximity to the large temples. They are generally erected on a raised terrace, rectangular on plan and nearly twice as long as it was wide, approached by a flight of steps opposite the entrance; they consist of oblong buildings, usually divided by a wall into two chambers, and surrounded on all sides by a colonnade composed of circular columns or square piers placed at intervals, and the whole is roofed in. A dwarf wall is fre-

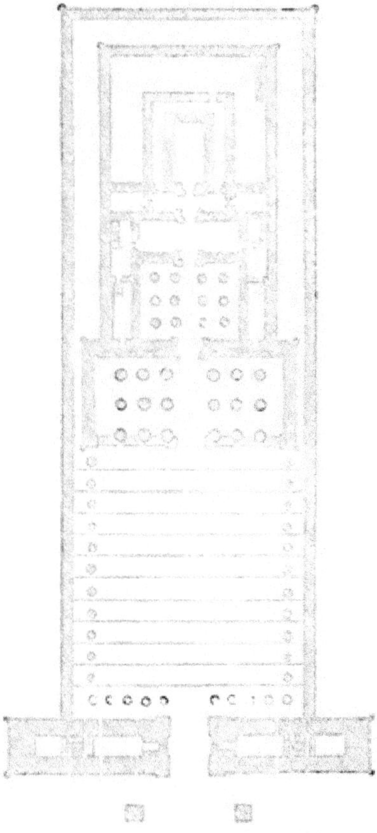

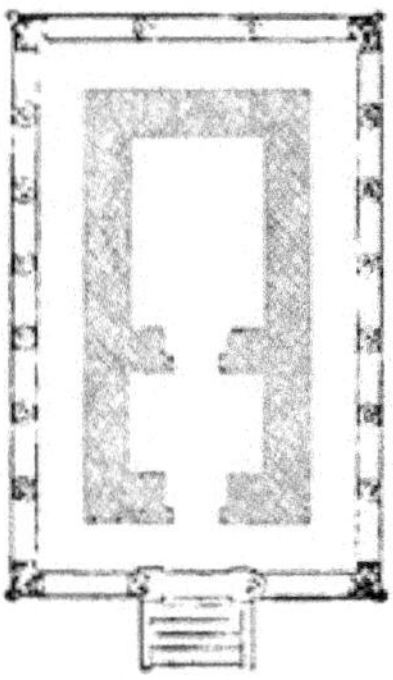

FIG. 20.—PLAN OF ONE OF THE MAMMISI AT EDFOU.

FIG. 19.—PLAN OF THE TEMPLE AT EDFOU.

quently found between the piers and columns, about half the height of the shaft. These temples differ from the larger ones in having their outer walls perpendicular. Fig. 20

is a plan of one of these small temples, and no one can fail to remark the striking likeness to some of the Greek temples; there can indeed be little doubt that this nation borrowed the peristylar form of its temples from the Ancient Egyptians.

Although no rock-cut temples have been discovered in Egypt proper, Nubia is very rich in such remains. The arrangement of these temples hewn out of the rock is closely analogous to that of the detached ones. Figs. 21 and 22 show a plan and section of the

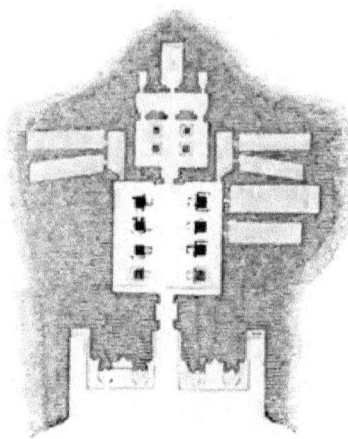

FIG. 21.—GROUND-PLAN OF THE ROCK-CUT TEMPLE AT IPSAMBOUL.

FIG. 22.—SECTION OF THE ROCK-CUT TEMPLE AT IPSAMBOUL.

largest of the rock-cut temples at Ipsamboul, which
consists of two extensive courts, with smaller chambers
beyond, all connected by galleries. The roof of the large
court is supported by eight huge piers, the faces of which
are sculptured into the form of standing colossi, and the
entrance is adorned by four splendid seated colossi,
68 ft. 6 in. high. As was the case with the detached
temples, it will be noticed that the height of the
various chambers decreases towards the extremity of the
excavation.

The constructional system pursued by the Egyptians,
which consisted in roofing over spaces with large horizontal
blocks of stone, led of necessity to a columnar arrangement
in the interiors, as it was impossible to cover large areas
without frequent upright supports. Hence the column
became the chief means of obtaining effect, and the varieties
of form which it exhibits are very numerous. The earliest
form is that at Beni-Hassan, which has already been noticed
as the prototype of the Doric order. Figs. 23 and 24 are
views of two columns of a type more commonly employed.
In these the sculptors appear to have intended as closely
as possible the forms of the plant-world around them, as
is shown in Fig. 23, which represents a bundle of reeds
or lotus stalks, and is the earliest type known of the lotus
column, which was afterwards developed into a number of
forms, one of which will be observed on turning to our
section of the Hypostyle Hall at Karnak (Fig. 17), as
employed for the lateral columns. The stalks are bound
round with several belts, and the capital is formed by
the slightly bulging unopened bud of the flower, above
which is a small abacus with the architrave resting upon
it : the base is nothing but a low circular plinth. The
square piers also have frequently a lotus bud carved on
them. At the bottom of the shaft is frequently found

a decoration imitated from the sheath of leaves from which the plant springs. As a further development of this capital we have the opened lotus flower of a very graceful bell-like shape, ornamented with a similar sheath-like decoration to that at the base of the shaft (Fig. 24). This decoration was originally painted only, not sculptured,

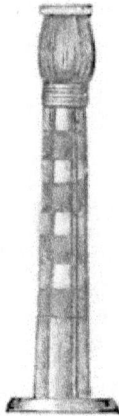

FIG. 23.—EGYPTIAN COLUMN
WITH LOTUS BUD CAPITAL.

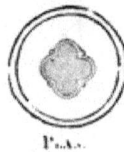

FIG. 24.—EGYPTIAN COLUMN
WITH LOTUS FLOWER CAPITAL.

but at a later period we find these sheaths and buds worked in stone. Even more graceful is the palm capital, which also had its leading lines of decoration painted on it at first (Fig. 25), and afterwards sculptured (Fig. 26). At a later period of the style we find the plant forms abandoned, and capitals were formed of a fantastic combination of the head

D

of Isis with a pylon resting upon it (Fig. 27). Considerable ingenuity was exercised in adapting the capitals of the columns to the positions in which they were placed : thus in the hypostyle halls, the lofty central row of columns generally had capitals of the form shown in Fig. 24, as the light here was sufficient to illuminate thoroughly the underside of the overhanging bell; but those columns which were farther removed from the light had their

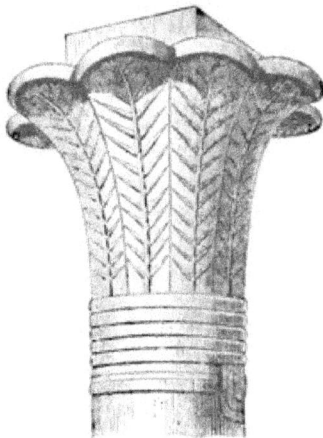

FIG. 25.—PALM CAPITAL.   FIG. 26.—SCULPTURED CAPITAL.

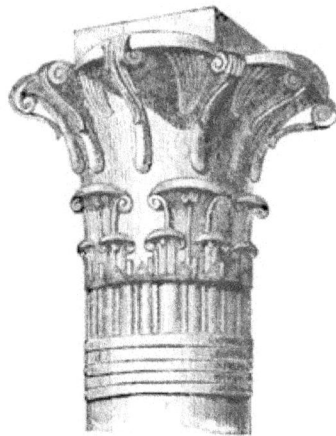

capitals of the unopened bud form, which was narrower at the top than at bottom. In one part of the temple at Karnak is found a very curious capital resembling the open lotus flower inverted. The proportion which the height of Egyptian columns bears to their diameter differs so much in various cases that there was evidently no regular standard adhered to, but as a general rule they have a heavy and massive character. The wall-paintings of the Egyptian buildings show many curious forms of columns (Fig. 28), but we have no reason for thinking that these fantastic shapes were really executed in stone.

Almost the only sculptured ornaments worked on the exteriors of buildings were the curious astragal or bead at all the angles, and the cornice, which consisted of a very large cavetto, or hollow moulding, surmounted by a fillet.

FIG. 27.—ISIS CAPITAL FROM DENDERAH.

FIG. 28.—FANCIFUL COLUMN FROM PAINTED DECORATION AT THEBES.

These features are almost invariable from the earliest to the latest period of the style. This cavetto was generally enriched, over the doorways, with an ornament representing a circular boss with a wing at each side of it. (Fig. 29).

One other feature of Egyptian architecture which was peculiar to it must be mentioned; namely, the obelisk.

Obelisks were nearly always erected in pairs in front
of the pylons of the temples, and added to the dignity of
the entrance.   They were invariably monoliths, slightly
tapering in outline, carved with the most perfect accuracy;
they must have existed originally in very large numbers.
Not a few of these have been transported to Europe, and
at least twelve are standing in Rome, one is in Paris, and
one in London.

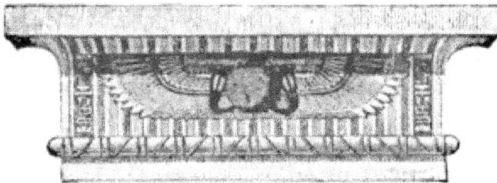

FIG. 29.—CROWNING CORNICE AND BEAD.

The most striking features, and the most artistic, in the
decoration of Egyptian buildings, are the mural paintings
and sculptured pictures, which are found in the most lavish
profusion, and which exhibit the highest skill in conventionalising the human figure and other objects.*   Tombs
and temples, columns and obelisks are completely covered
with graphic representations of peaceful home pursuits,
warlike expeditions and battle scenes, and—though not
till a late period—descriptions of ritual and mythological
delineations of the supposed spirit-world which the soul
has entered after death.   These pictures, together with the

* Conventionalising may be described as representing a part only of the
visible qualities or features of an object, omitting the remainder or very
slightly indicating them.   A black silhouette portrait is an extreme
instance of convention, as it displays absolutely nothing but the outline
of a profile.   For decorative purposes it is almost always necessary to
conventionalise to a greater or less extent whatever is represented.

hieroglyphic inscriptions—which are in themselves a series of pictures—not only relieve the bare wall surface, but, what is far more important, enable us to realise the kind of existence which was led by this ancient people; and as in nearly every case the cartouche (or symbol representing the name) of the monarch under whose reign the building was erected was added, we should be able to fix the dates of the buildings with exactness, were the chronology of the kings made out beyond doubt.

The following description of the manner in which the Egyptian paintings and sculptures were executed—from the pen of Owen Jones—will be read with interest :— "The wall was first chiselled as smooth as possible, the imperfections of the stone were filled up with cement or plaster, and the whole was rubbed smooth and covered with a coloured wash; lines were then ruled perpendicularly and horizontally with red colour, forming squares all over the wall corresponding with the proportions of the figure to be drawn upon it. The subjects of the painting and of the hieroglyphics were then drawn on the wall with a red line, most probably by the priest or chief scribe, or by some inferior artist, from a document divided into similar squares; then came the chief artist, who went over every figure and hieroglyphic with a black line, and a firm and steady hand, giving expression to each curve, deviating here and confirming there the red line. The line thus traced was then followed by the sculptor. The next process was to paint the figure in the prescribed colours."

Although Egyptian architecture was essentially a trabeated style,—that is to say, a style in which beams or lintels were usually employed to cover openings,—there is strong ground for the belief that the builders of that time

were acquainted with the nature of the arch. Dr. Birch
mentions a rudimentary arch of the time of the fifth
dynasty: at Abydos there are also remains of vaulted
tombs of the sixth dynasty; and in a tomb in the neighbourhood of the Pyramids there is an elementary arch of
three stones surmounted by a true arch constructed in four
courses. The probability is that true brick arches were
built at a very early period, but in the construction of
their tombs, where heavy masses of superincumbent
masonry or rock had to be supported, the Egyptians
seem to have been afraid to risk even the remote possi-
bility of their arches decaying; and hence, even when
they preserved the form of the arch in masonry, they
constructed it with horizontal courses of stone projecting
one over the other, and then cut away the lower angles.
One dominating idea seems to have influenced them in
the whole of their work—*esto perpetua* was their motto;
and though they have been excelled by later peoples in
grace and beauty, it is a question whether they have ever
been surpassed in the skill with which they adapted their
means to the end which they always kept in view.

ANALYSIS OF BUILDINGS.

Plan.

Floor (technically *Plan*).—The early rock-cut tombs
were, of course, only capable of producing internal effects;
their floor presents a series of halls and galleries, varying in size and shape, leading one out of the other, and
intended by their contrast or combination to produce
architectural effect. To this was added in the later rock-
cut tombs a façade to be seen directly in front. Much
the same account can be given of the disposition of the

built temples. They possess one front, which the spectator approaches, and they are disposed so as to produce varied and impressive interiors, but not to give rise to external display. The supports, such as walls, columns, piers, are all very massive and very close together, so that the only wide open spaces are courtyards.

The circle, or octagon, or other polygonal forms do not appear in the plans of Egyptian buildings; but though all the lines are straight, there is a good deal of irregularity in spacing, walls which face one another are not always parallel, and angles which appear to be right angles very often are not so.

The later buildings extend over much space. The adjuncts to these buildings, especially the avenues of sphinxes, are planned so as to produce an air of stately grandeur, and in them some degree of external effect is aimed at.

Walls.

The walls are uniformly thick, and often of granite or of stone, though brick is also met with; *e.g.* some of the smaller pyramids are built entirely of brick. In all probability the walls of domestic buildings were to a great extent of brick, and less thick than those of the temples; hence they have all disappeared.

The surface of walls, even when of granite, was usually plastered with a thin fine plaster, which was covered by the profuse decoration in colour already alluded to.

The walls of the propylons tapered from the base towards the top, and the same thing sometimes occurred in other walls. In almost all cases the stone walls are built of very large blocks, and they show an unrivalled skill in masonry.

Roofs.

The roofing which remains is executed entirely in stone, but not arched or vaulted. The rock-cut tombs, however, as has been stated, contain ceilings of an arched shape, and in some cases forms which seem to be an imitation of timber roofing. The roofing of the Hypostyle Hall at Karnak provides an arrangement for admitting light very similar to the clerestory of Gothic cathedrals.

Openings.

The openings were all covered by a stone lintel, and consequently were uniformly square-headed. The interspaces between columns were similarly covered, and hence Egyptian architecture has been, and correctly, classed as the first among the styles of trabeated architecture. Window-openings seldom occur.

Columns.

The columns have been already described to some extent. They are almost always circular in plan, but the shaft is sometimes channelled. They are for the most part of sturdy proportions, but great grace and elegance are shown in the profile given to shafts and capitals. The design of the capitals especially is full of variety, and admirably adapts forms obtained from the vegetable kingdom. The general effect of the Egyptian column, wherever it is used, is that it appears to have, as it really has, a great deal more strength than is required. The fact that the abacus (the square block of stone introduced between the moulded part of the capital and what it carries) is often smaller in width than the diameter of the column aids very much to produce this effect.

Ornaments.

Mouldings are very rarely employed; in fact, the large bead running up the angles of the pylons, &c., and a heavy hollow moulding doing duty as a cornice, are all that are usually met with. Sculpture and carving occur occasionally, and are freely introduced in later works, where we sometimes find statues incorporated into the design of the fronts of temples. Decoration in colour, in the shape of hieroglyphic inscriptions and paintings of all sorts, was profusely employed (Figs. 27–30), and is executed with a truth of drawing and a beauty of colouring that have never been surpassed. As has been pointed out, almost every object drawn is partly conventionalised, in the most skilful manner, so as to make it fit its place as a piece of a decorative system.

Architectural Character.

This is gloomy, and to a certain extent forbidding, owing to the heavy walls and piers and columns, and the great masses supported by them; but when in its freshness and quite uninjured by decay or violence, the exquisite colouring of the walls and ceilings and columns must have added a great deal of beauty: this must have very much diminished the oppressive effect inseparable from such massive construction and from the gloomy darkness of many portions of the buildings. It is also noteworthy that the expenditure of materials and labour is greater in proportion to the effect attained than in any other style. The pyramids are the most conspicuous example of this prodigality. Before condemning this as a defect in the style, it must be remembered that a stability which should defy enemies, earthquakes, and the tooth of time, was far more aimed at than architectural character; and that, had any mode of construction less lavish of material, and less perfect in workmanship, been adopted, the buildings of Egypt might have all disappeared ere this.

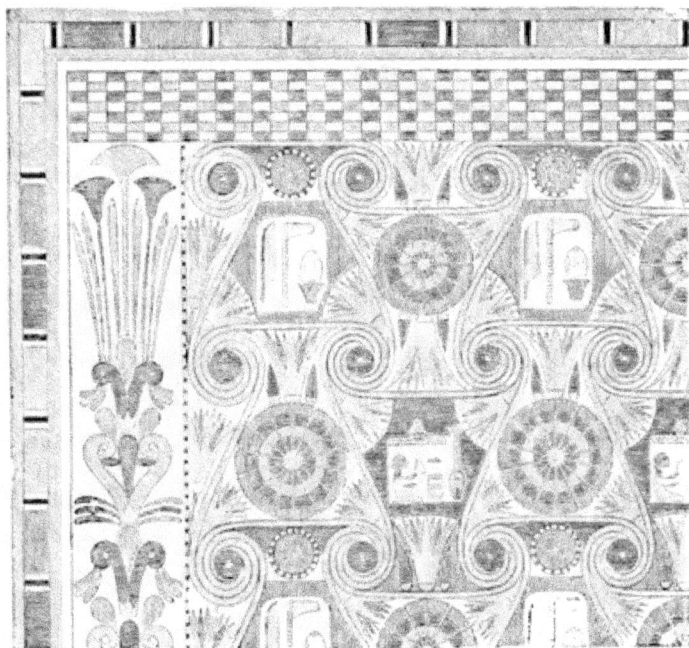

FIG. 30.—PAINTED DECORATION FROM THEBES.

FIG. 31.—SCULPTURED ORNAMENT AT NINEVEH.

CHAPTER III.

WEST ASIATIC ARCHITECTURE.

THE architectural styles of the ancient nations which ruled over the countries of Western Asia watered by the Tigris and the Euphrates, from a period about 2200 B.C. down to 330 B.C., are so intimately connected one with another, and so dependent one upon the other, that it is almost impossible to attempt an accurate discrimination between the Babylonian or ancient Chaldæan, the Assyrian and the Persian. A more intelligible idea of the architecture of this long period will be gained by regarding the three styles as modifications and developments of one original style, than by endeavouring to separate them.[*] Their sequence can, however, be accurately determined. First comes the old Chaldæan period, next the Assyrian, during which the great city of Nineveh was built, and

---

* In any such endeavour we should be met by the further difficulty, that the writers of antiquity differ widely in the precise limits which they give to the Assyrian Kingdom. Some make it include Babylon, other writers say that it was bounded on the south by Babylon, and altogether the greatest confusion exists in the accounts that have come down to us.

finally the Persian, after Cyrus had subdued the older monarchies; and remains exist of all these periods. As to the origin of the Chaldæan Kingdom, however, all is obscure; and the earliest date which can be fixed with the slightest approach to probability is 2234 B.C., when Nimrod is supposed to have founded the old Chaldæan dynasty. This seems to have lasted about 700 years, and was then overthrown by a conquering nation of which no record or even tradition remains, the next two and a half centuries being a complete blank till the rise of the great Assyrian Monarchy about 1290 B.C., which lasted till its destruction by Cyrus about 538 B.C. The Persian Monarchy then endured till the death of Alexander the Great, in 333 B.C., after which great confusion arose, the empire being broken up among his generals and rapidly falling to pieces.

It is only within a comparatively recent period that we have had any knowledge of the architecture of these countries; but the explorations of M. Botta, commenced in 1843 and continued by M. Place, and those of Mr. (now Sir A. H.) Layard in 1845, combined with the successful attempts of Prof. Grotefend, Prof. Lassen, and Col. Rawlinson at deciphering the cuneiform inscriptions, have disclosed a new world to the architectural student, without which some of the developments of Greek architecture must have remained obscure. The authentic remains of buildings of the early Chaldæan period are too few and in too ruinous a condition to allow of a reproduction of their architectural features with any certainty. The buildings, whether palaces or temples, appear to have been constructed on terraces, and to have been several storeys in height ; and in one instance, at Mugheyr, the walls sloped inwards in a similar manner to those of Egyptian buildings, a peculiarity which is not met with in other examples of West Asiatic architecture. The materials employed were bricks, both sun-dried and kiln-burnt, which seem to have been coated with a vitreous enamel for purposes of interior decoration. Fragments of carved limestone were discovered by Sir A. H. Layard, but the fact that the fragments found have been so few ought not to lead us too hastily to the conclusion that stone was not used as facing for architectural purposes, as after the buildings became ruined the stone would eagerly be sought for and carried away before the brickwork was touched. Bitumen seems to have been employed as a cement. Although original buildings of this era cannot be found, it has been shown that in all probability we have, in a building of a later date—the Birs-i-Nimrud—a type of the old Babylonian temple. This in its general disposition must have resembled that of the Tomb of Cyrus, described and figured later on, though on a vastly larger scale. The lowest storey appears to have been an exact square of 272 ft.; each of the higher storeys was 42 ft. less horizontally than the one below it, and was placed 30 ft. back from the front of the storey below it, but equidistant from the two sides, where the platforms were 21 ft. wide. The three upper storeys were 45 ft. in heightal together, the two below these were 26 ft. each, and the height of the lowest is uncertain. The topmost storey probably had a tower on it which enclosed the shrine of the temple. This edifice was for a long time a bone of contention among savants, but Colonel Rawlinson's investigations have brought to light the fact that it was a temple dedicated to the seven heavenly spheres, viz. Saturn, Jupiter, Mars, the Sun, Venus, Mercury, and the Moon, in the order given, starting from the bottom. Access to the

various platforms was obtained by stairs, and the whole
building was surrounded by a walled enclosure. From
remains found at Wurkha we may gather that the walls
of the buildings of this period were covered with elaborate
plaster ornaments, and that a lavish use was made of
colour in their decoration.

Of the later Assyrian period several ruins of buildings
believed to be palaces have been excavated, of which the
large palace at Khorsabad, the old name of which was
Hisir-Sargon, now a small village between 10 and 11 miles
north-east of Nineveh, has been the most completely
explored, and this consequently is the best adapted to explain the general plan of an Assyrian edifice. M. Botta,
when French Consul at Mosul, and M Victor Place con-
ducted these explorations, and the following details are
taken from their works. Like all other Assyrian palaces,
this was reared on a huge artificial mound, the labour
of forming which must have been enormous. The reason
for the construction of these mounds is not far to seek.
Just as the chiefs of a mountainous country choose the
loftiest peaks for their castles, so in Assyria, which
was a very flat country, the extra defensive strength
of elevated buildings was clearly appreciated; and as
these absolute monarchs ruled over a teeming popula-
tion and had a very large number of slaves, and only had
to direct their taskmasters to impress labour whenever
they wanted it, no difficulty existed in forming elevated
platforms for their palaces. These were frequently close
to a river, and it is by no means improbable that this was
turned into the excavation from which the earth for the
mound was taken, and thus formed a lake or moat as an
additional defence. A further reason for these terraces
may be found in the fact that in a hot climate buildings

erected some 20 or 30 ft. above the level of the plain catch the breezes much more quickly than lower edifices. In the case of Khorsabad the terrace was made of sun-dried bricks, about 15·7 in. square and 2 in. thick. These bricks were made of the most carefully prepared clay. The terrace was faced by a retaining wall of coursed masonry, nearly 10 ft. in thickness. On this terrace the palace was built, and it consisted of a series of open courts arranged unsymmetrically, surrounded by state or private apartments, storehouses, stables, &c. Great care seems to have been exercised in the accurate orientation of the building, but in rather a peculiar manner. Instead of any one façade of the building facing due north, the corners face exactly towards the four points of the compass. The courts were all entered by magnificent portals flanked by gigantic figures, and were approached by flights of steps. Fig. 32 is a plan of the palace of Khorsabad, which was placed close to the boundary of the city; in fact it was partly outside the city wall proper, though surrounded by a wall of its own. The grand south-east portals or propylæa were adorned with huge human-headed bulls and gigantic figures, and gave access to a large court, 315 ft. by 280 ft., on the east side of which are the stables and out-houses, and on the west side the metal stores. On the north of this court, though not approached directly from it, was the Seraglio (not to be confounded with the Harem), the grand entrance to which was from a second large court, access to which was obtained from a roadway sloping up from the city. The portals to this portion of the palace were also adorned with human-headed bulls. From the second court a vaulted passage gave access to the state apartments, which appear to have had a direct view across the open country, and

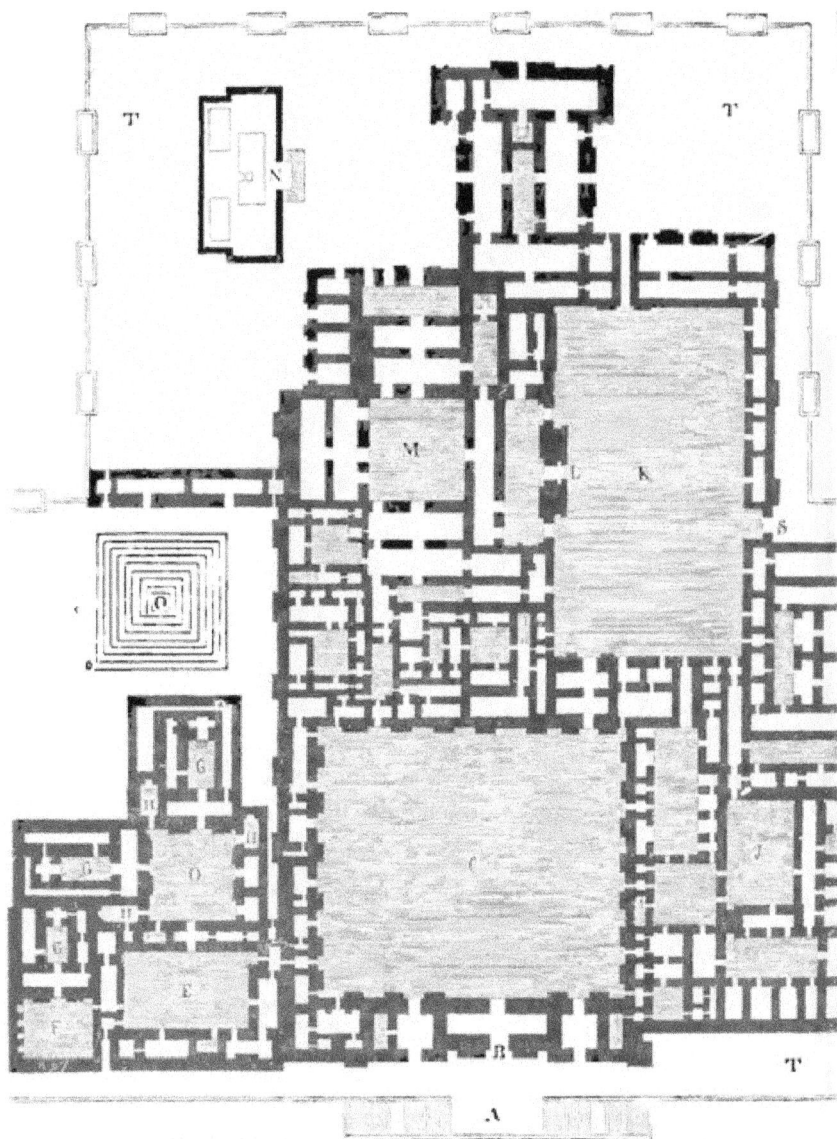

FIG. 32.—PALACE OF KHORSABAD. BUILT BY KING SARGON ABOUT 710 B.C.

A, Steps. B, Chief portal. C, Chief entrance-court. D—H, Women's apartments (Harem). J, Centre court of building. K, Chief court of royal residence. L, Portal with carved bulls as guards. M, Centre court of royal residence. N, Temple (?). O, Pyramid of steps. S, Entrance to chief court. T, Plan of terraces with wall and towers.

were quite outside the city walls. The Harem has been
excavated; it stood just outside the palace proper, behind
the metal stores. The remains of an observatory exist,
and the outlines of what is supposed to have been a
temple have also been unearthed, so that we have here a
complete plan of the palace. Altogether 31 courts and
198 chambers have been discovered.

It will be noticed that great disproportion exists between
the length of the various apartments and their breadth, none
being more than 40 ft. wide; and it is probable that this
was owing to structural necessities, the Assyrian builders
finding it impossible, with the materials at their disposal,
to cover wider spaces than this. The walls of this palace
vary from 5 to 15 ft. in thickness, and are composed of
sun-dried bricks, faced in the principal courts and state
apartments with slabs of alabaster or Mosul gypsum to a
height of from 9 to 12 ft., above which kiln-burnt bricks
were used. The alabaster slabs were held together by
iron, copper, or wooden cramps or plugs, and were covered
with sculptured pictures representing scenes of peace and
war, from which, as was the case with the Egyptian
remains, we are able to reconstruct for ourselves the daily
life of the monarchs of those early times. Above the
alabaster slabs plastered decorations were used; in some
cases painted frescoes have been found, or mosaics formed
with enamelled bricks of various colours. In the out-
buildings and the more retired rooms of the palace, the
alabaster slabs were omitted, and plaster decorations used,
from the ground upwards. The researches of MM. Botta
and Place have shown that colour was used with a lavish-
ness quite foreign to our notions, as the alabaster statues
as well as the plaster enrichments were coloured. M. Place
says that in no case were the plain bricks allowed to face

E

the walls of an apartment, the joint being always con-
cealed either by colour or plaster : in fact, he remarks that
after a time, if he found walls standing showing the brickwork joints, he invariably searched with success among
the débris of the chamber for remains of the sculptured
decorations which had been used to face the walls.

Not the least interesting of these discoveries was that
of the drains under the palace, portions of which were
in very good preservation ; and all were vaulted, so that
there can be no doubt whatever that the Assyrians were
acquainted with the use of the arch. This was further
proved by the discovery by M. Place of the great arched
gates of the city itself, with an archivolt of coloured
enamelled bricks forming various patterns, with a semicircular arch springing from plain jambs. Extreme care
was taken by the Assyrian builders in laying the pavements to ensure their being perfectly level : first a layer
of kiln-burnt bricks was laid on the ordinary sun-dried
bricks forming the terrace ; then came a layer of fine sand,
upon which the bricks or slabs of the pavement proper
were laid, forming in many cases an elegant pattern (see
Fig. 33).

Great difference of opinion exists as to the manner in
which the various apartments of the palace were lighted.
M. Place suggests that the rooms were all vaulted on the
inside, and the spandrels filled in with earth afterwards to
form perfectly flat roofs, and he gives a restoration of the
building on such an arrangement ; but if he is correct,
it is impossible to see how any light at all can have pene-
trated into the interior of many of the apartments, and
as these apartments are decorated with a profusion of
paintings it is very difficult to believe that artificial light
alone was used in them. M. Place thinks, however, that

in some cylindrical terra-cotta vessels which he found he
has hit upon a species of skylight which passed completely
through the vault over the rooms, and thus admitted the
light from above. This, however, can hardly be considered
as settled yet. Mr. Fergusson, on the other hand, suggests
that the thick main walls were carried to a height of about
18 or 19 ft., and that above this were two rows of dwarf
columns, one on the inner and the other on the outer edge

Fig. 33.—Pavement from Khoyunjik.

of the wall, these columns supporting a flat terrace roof, and
the walls thus forming galleries all round the apartments.
Then to cover the space occupied by the apartments themselves it is necessary to assume the existence of rows of
columns, the capitals of which were at the same level as
those of the dwarf columns on the walls. Where one
apartment is surrounded on all sides by others, the roof
over it may have been carried up to a higher level, forming

a sort of clerestory. This theory no doubt accounts for many things which are very hard to explain otherwise, and derives very strong support from the analogy of Persepolis, where slender stone columns exist. Such columns of cedar wood would add enormously to the magnificence and grandeur of the building; and if, as seems likely, most of these Assyrian palaces were destroyed by fire, the absence of the remains of columns offers no difficulty. On the other hand, in many parts of the palace of Khorsabad no trace of fire remains, and yet here no suggestion of detached columns can be found, and, moreover, it is extremely difficult to arrange columns symmetrically in the various apartments so that doorways are not interfered with. There is also another difficulty, viz. that if the building called the Harem at Khorsabad was built in this way, the apartments would have been open to the view of any one ascending the lofty building called the observatory. It is quite possible that further explorations may tend to elucidate this difficult question of roofing, but at present all that can be said is that none of these theories that have been put forward is wholly satisfactory.

As no columns at all exist, we cannot say what capitals were employed, but it is probable that those of Persepolis, which will be shortly described, were copied from an earlier wooden form, which may have been that used by the Assyrian builders. There is, however, capping the terrace on which the temple was erected at Khorsabad, a good example of an Assyrian cornice, which is very similar indeed to the forms found in Egypt, and some of the sculptured bas-reliefs which have been discovered depict rude copies of Assyrian buildings drawn by the people themselves; and it is most interesting to notice that just as we found in the Egyptian style the proto-Doric column, so in the Assyrian we find the proto-Ionic (Figs. 34, 34A), and possibly also the proto-Corinthian (Fig. 34B).

The third branch of West Asiatic architecture is the Persian, which was developed after Cyrus had conquered the older monarchies, and which attained its greatest magnificence under Darius and Xerxes. The Persians were originally a brave and hardy race inhabiting the mountainous region south of Media, which slopes down to the Persian Gulf. Until the time of Cyrus, who was the founder

FIG. 34A.—PROTO-IONIC
CAPITAL FROM ASSYRIAN
SCULPTURE.

FIG. 34.—PROTO-
IONIC COLUMN.

FIG. 34B.—PROTO-
CORINTHIAN CAPITAL
FROM ASSYRIAN
SCULPTURE.

of the great kingdom of Persia, they inhabited small towns, had no architecture, and were simple barbarians. But after Cyrus had vanquished the wealthy and luxurious Assyrian monarchs, and his warriors had seen and wondered at the opulence and splendour of the Assyrian palaces, it was natural that his successors should strive to emulate for themselves the display of their vassals. Therefore, having no indigenous style to fall back upon, the artisans who were summoned to build the tomb of the founder of the monarchy and the palaces of his successors, simply copied the forms with which they were acquainted. Fortunately, the sites for the new palaces were in a locality where building stone was good and abundant, and the presence of this material had a modifying effect upon the architecture.

The best known of the remains which date as far back as the earlier Persian dynasties is the so-called tomb of Cyrus at Pasargadæ, near Murghab (Fig. 35). This may

FIG 35 – TOMB OF CYRUS.

be looked upon as a model in white marble of an old Chaldæan temple, such as the Birs-i-Nimrud. There are the same platforms diminishing in area as the top is approached, and on the topmost platform is a small cella or temple with a gabled stone roof, which probably originally contained the sarcophagus. It is, however, at Persepolis, the real capital of the later Persian kings, whose grandeur and wealth were such that Alexander is said to have found there treasure to the amount of thirty millions of pounds sterling, that we find the most magnificent series of ruins. These were carefully measured and drawn by Baron Texier in 1835, and his work and that of MM. Flandrin and Coste are those from which the best information on this subject can be obtained.

Persepolis is about 35 miles north-east of Shiraz, close to the main highway to Ispahan, at the foot of the mountain range which bounds the extensive plain of Nurdusht. The modern inhabitants of the district call the ruins Takht-i-Jamshid (or the building of Jamshid), but the inscriptions that have been deciphered prove that Darius and Xerxes were the chief builders. Just as was the case with the Assyrian ruins, these stand on an immense platform which rises perpendicularly from the plain and abuts in the rear against the mountain range. Instead, however, of this platform being raised artificially, it was cut out of the rock, and levelled into a series of terraces, on which the buildings were erected. The platform, whose length from north to south is about 1582 ft., and breadth from east to west about 938 ft., is approached from the plain by a magnificent double staircase of black marble, of very easy rise, not more than 4 in. each step. Its general height above the level of the plain was originally 34 ft. 9 in. The retaining wall of the platform is not straight, but has in it 40 breaks or set-offs of unequal dimensions. At the top of the staircase are the remains of a building with four columns in the centre and with large portals both back and front, each of which is adorned with gigantic bulls, strikingly resembling those found at Khorsabad. Those in the front have no wings, but those in the rear have wings and human heads. It has been suggested that these are the ruins of one of those large covered gates

frequently mentioned in the Bible, under the shelter of
which business was transacted, and which probably formed
the entrance to the whole range of courts and buildings.
After passing through this gateway and turning south-

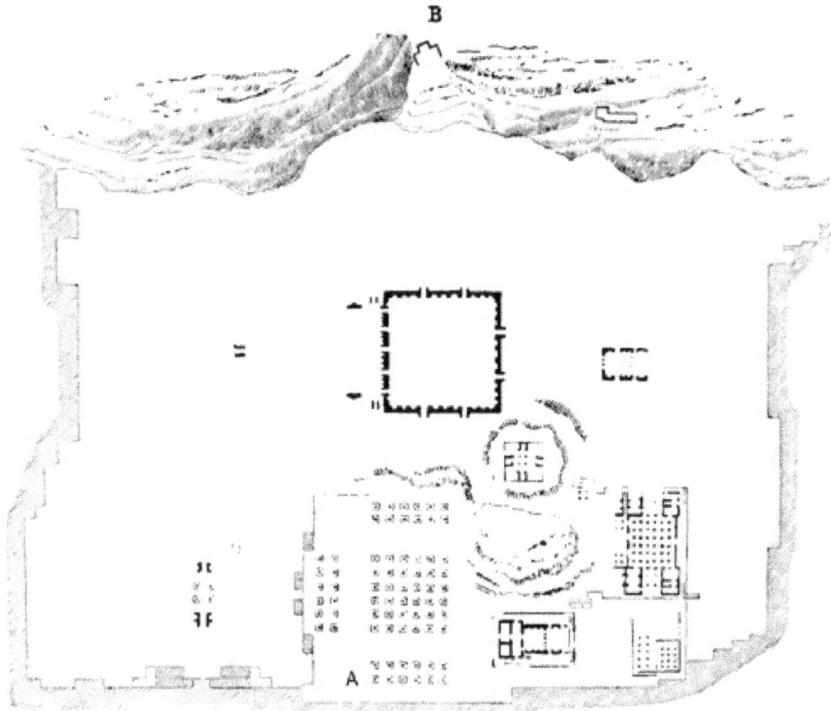

FIG. 35A.—GENERAL PLAN OF THE BUILDINGS AT PERSEPOLIS.

wards, at a distance of 177 feet from it, another terrace
is reached, 9 ft. 2 in. higher than the first one. This terrace also is approached by four flights of steps profusely
decorated with sculptured bas-reliefs, and on it are the
remains of the Chehil Minar, the grand hexastyle Hall of

Xerxes, which must have been one of the most magnificent buildings of ancient times. This building is marked A on the general plan. It consisted of a central court, containing thirty-six columns, the distance from centre to centre of the outside columns being 142 ft. 8 in. This court was surrounded by walls, of which nothing now remains but the jambs of three of the doorways. On three sides of this court, to the north, east and west, were porticoes of twelve columns each, precisely in a line with those of the central court, the distance from centre to centre of the columns being 28 ft. 6 in. These columns, both in their proportions and shape, suggest an imitation of timber construction. On the south the court was probably terminated by a wall, and Mr. Fergusson suggests that the corners between the porticoes were filled up with small chambers. The most striking feature of this hall or palace must have been its loftiness, the height of the columns varying from 63 ft. 8 in. to 64 feet from bottom of base to top of capital. The shafts were slightly tapering and had 48 flutings, and were 4 ft. 6 in. in diameter in the upper part. The bases of the columns show hardly any variations, and consist of a series of mouldings such as is shown in Fig. 36; the lowest part of this moulded base is enriched with leaves, and rests on a low circular plinth at the bottom: the total height of the base averages 5 feet. The capitals show considerable variations. Those of the east and west porticoes represent the heads and fore part of the bodies of two bulls* placed directly on the shaft back to back, with their forelegs doubled under them,

---

* As a matter of fact there is a marked distinction between the heads of the animals of the east and west porticoes : those of the west are undoubtedly bulls, but those of the east are grotesque mythological creatures somewhat resembling the fabled unicorn.

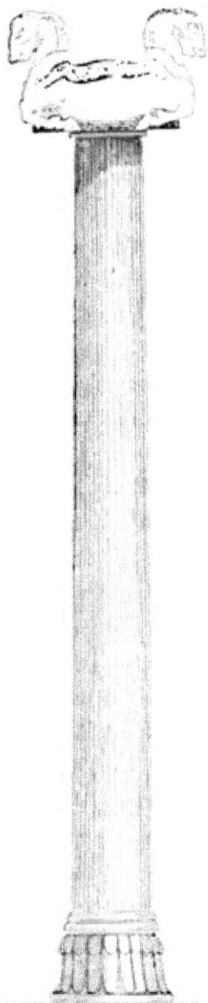

FIG. 35. — COLUMN FROM PERSE-
POLIS, EAST AND WEST PORTICOES.

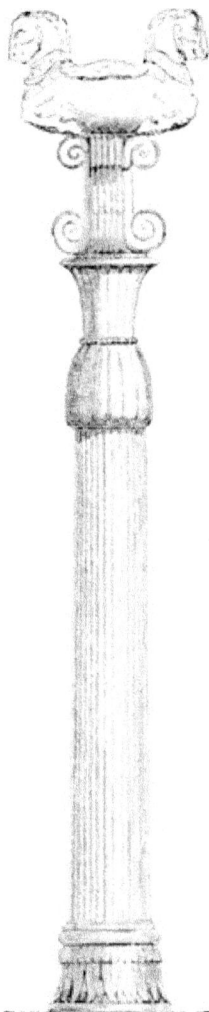

FIG. 36. — COLUMN FROM
PERSEPOLIS, NORTH PORTICO.

the feet resting on the shaft and the knees projecting;
the total height of these capitals is 7 ft. 4 in. Between
the necks of the bulls rested the wooden girder which
supported the cross-bearers of the roof. In the north
portico and, so far as can be ascertained, in the central
court, the shaft of the column was much shorter, and
supported a fantastic elongated capital, consisting of a
sort of inverted cup, supporting an elegant shape much
resembling the Egyptian palm-leaf capital, above which,
on all the four sides, are double spirals resembling the
ornaments of the Greek Ionic capital known as volutes, but
placed perpendicularly, and not, as in the Ionic capital,
horizontally. These volutes again may have supported
double bulls, which would make the total height of the
columns the same as those of the east and west porticoes.
The doorways have cornices enriched with leaves, similar
to those found at Khorsabad, which have already been
noticed as bearing a decided resemblance to the Egyptian
doorways.

On other terraces, slightly raised above the main platform, exist the remains, in a more or less ruined condition,
of numerous other courts and halls, one of which has no
less than one hundred columns to support its roof, but the
height of this building was much inferior to that of the
Chehil Minar. The existence of these columns leaves no
doubt that these buildings were covered with flat roofs;
and that over part of them was a raised talar or prayer-
platform is rendered probable from the introduction of such
a feature into the sculptured representation of a palace
façade which forms the entrance to the rock-tomb of
Darius, which was cut out of the mountain at the back
of the terrace of Persepolis. The position of this tomb
on the general plan is marked B, and Fig. 37 is a view

of the entrance, which was probably intended as a copy
of one of the halls. All the walls of the palaces were
profusely decorated with sculptured pictures, and various
indications occur which induce the belief that painting
was used to decorate those portions of the walls that were
not faced with sculptured slabs.

FIG. 32.—THE ROCK-CUT TOMB OF DARIUS.

The superior lightness and elegance of the Persepolitan
ruins to those of an earlier epoch will not fail to be noticed,
but there is still a certain amount of barbaric clumsiness
discernible, and it is not till we come to Greek architecture that we see how an innate genius for art and

beauty, such as was possessed by that people, could cull from previous styles everything capable of being used with effect, and discard or prune off all the unnecessary exuberances of those styles which offend a critically artistic taste.

ANALYSIS OF BUILDINGS.

Plan.

The floor-space of a great Assyrian or Medo-Persian building was laid out on a plan quite distinct from that of an Egyptian temple; for the rooms are almost always grouped round quadrangles. The buildings are also placed on terraces, and no doubt would secure external as well as internal effects, to which the imposing flights of stairs provided would largely contribute. We find in Assyrian palaces, halls comparatively narrow in proportion to their great length, but still so wide that the roofing of them must have been a serious business, and we find them arranged side by side, often three deep. In the Persian buildings, halls nearly square on plan, and filled by a multitude of columns, occur frequently. In the plan of detached buildings like the Birs-i-Nimrud, we are reminded of the pyramids of Egypt, which no doubt suggested the idea of pyramidal monuments to all subsequent building peoples.

Walls.

The magnificently worked granite and stones of Egypt give place to brick for the material of the walls, with the result that a far larger space could be covered with buildings by a given number of men in a given time, but of course the structures were far more liable to decay. Accordingly, sturdy as their walls are, we find them at the present day reduced to mere shapeless mounds, but of prodigious extent.

Roofs.

We can only judge of the roofs by inference, and it has already been stated that a difference of opinion exists respecting them. It appears most probable that a large proportion of the buildings must have been roofed by throwing timber beams from wall to wall and forming a thick platform of earth on them, and must have been lighted by some sort of clerestory. At any rate the stone roofs of the Egyptians seem to have been discarded, and with them the necessity for enormous columns and piers placed very close together. In some bas-reliefs, buildings with roofs of a domical shape are represented.

Openings.

Doorways are the openings chiefly met with, and it is not often that the super-structure, whether arch or lintel, remains, but it is clear that in some instances, at least, openings were arched. Great attention was paid to important doorways, and a large amount of magnificent sculpture was employed to enrich them.

Columns.

The columns most probably were of wood in Assyrian palaces. In some of the Persian ones they were of marble, but of a proportion and treatment which point to an imitation of forms suitable for wood. The bases and capitals of these slender shafts are beautiful in themselves, and very interesting as suggesting the source from which some of the forms in Greek architecture were derived; and on the bas-reliefs other architectural forms are represented which were afterwards used by the Greeks.

Ornaments.

Sculptured slabs, painted wall decorations, and terracotta ornamentation were used as enrichments of the walls. These slabs, which have become familiarly known through the attention roused by the discoveries of Sir A. H. Layard and the specimens sent by him to the British Museum, are objects of the deepest interest; so are the carved bulls from gateways. In the smaller and more purely ornamental decorations the honeysuckle, and other forms familiar to us from their subsequent adoption by Greek artists, are met with constantly, executed with great taste.

Architectural Character.

A character of lavish and ornate magnificence is the quality most strongly displayed by the architectural remains of Western Asia; and could we have beheld any one of the monuments before it was reduced to ruin, we should probably have seen this predominant to an extent of which it is almost impossible now to form an adequate idea.

FIG. 38.—SCULPTURED ORNAMENT AT ALLAHABAD.

CHAPTER IV.

ORIENTAL ARCHITECTURE.

Hindu Architecture.

H INDU architecture is not only unfamiliar but uncongenial to Western tastes; and as it has exercised no direct influence upon the later styles of Europe, it will be noticed in far less detail than the magnitude and importance of many Indian buildings which have been examined and measured during the last few years would otherwise claim, although the exuberant wealth of ornament exhibited in these buildings denotes an artistic genius of very high order, if somewhat uncultured and barbaric. As by far the largest number of Hindu buildings are of a date much later than the commencement of our era, a strict adherence to chronological sequence would scarcely allow the introduction of this style so early in the present volume; but we know that several centuries before Christ powerful kingdoms and wealthy cities existed in India; and as it seems clear also that in architecture and art, as well as in

manners and customs, hardly any change * has occurred
from remote antiquity, it appeared allowable, as well as
convenient, that the short description we have to offer
should precede rather than follow that of the classical
styles properly so called. Here, as always when we
attempt to penetrate farther back than a certain date,
all is obscure and mythical. We find lists of kings
and dynasties going back thousands of years before our
era, but nothing at all to enable us to judge how much of
this may be taken as solid fact. Mr. Fergusson believes
he has discovered in one date, viz. 3101 B.C., the first Aryan
settlement; but be this as it may, it is useless to look for
any architectural remains until after the death of Gotama
Buddha in 543 B.C.; in fact, it is very doubtful whether
remains can be authenticated until the reign of King
Asoka (B.C. 272 to B.C. 236), when Buddhism had spread
over almost the whole of the country, where it remained
the predominant cult until Brahmanism again asserted its
supremacy in the 14th century A.D.

The earliest, or among the earliest, architectural remains are the inscribed pillars called Lâts, which are
found in numerous localities, but have been almost always
overthrown. Many of these were erected by the above-
named Asoka: they were ornamented with bands and
mouldings separating the inscriptions, and crowned by a
sort of capital, which was generally in the form of an
animal. One very curious feature in these pillars is the

---

* It is not intended to imply that Hindustan has been without change
in her ruling dynasties. These have been continually changing; but the
remarkable fact is that, numerous as have been the nations that have
poured across the Indus attracted by " the wealth of Ind," there has been
no reflux, as it were: the various peoples, with their arts, religions, and
manners, have been swallowed up and assimilated, leaving but here and
there slight traces of their origin.

F

constant occurrence of a precise imitation of the well-known
honeysuckle ornament of the Greeks; this was probably
derived from the same source whence the Greeks obtained
it, namely Assyria. It is most probable that these pillars
served to ornament the approaches to some kind of sacred
enclosure or temple, of which, however, no remains have
been found.

Extremely early in date are some of the tumuli or topes
which exist in large numbers in various parts of India.

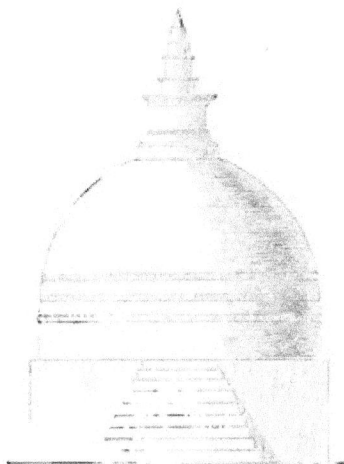

FIG. 39.—DAGOBA FROM CEYLON.

These are of two kinds,—the topes or stupas proper, which
were erected to commemorate some striking event or to
mark a sacred spot; and the dagobas, which were built
to cover the relics of Buddha himself or some Buddhist
saint. These topes consist of a slightly stilted hemispherical dome surmounting a substructure, circular in
plan, which forms a sort of terrace, access to which is
obtained by steps. The domical shape was, however, ex-

ternal only, as on the inside the masonry was almost solid,
a few small cavities only being left for the protection of
various jewels, &c. The dome was probably surmounted
by a pinnacle, as shown in Fig. 39. In the neighbourhood
of Bhilsa, in Central India, there are a large number of
these topes, of which the largest, that of Sanchi, measures
121 ft. in diameter and 55 ft. in height; it was erected
by King Asoka.

Two kinds of edifices which are not tombs remain,
the chaityas (temples or halls of assembly) and viharas
or monasteries, which were generally attached to the
chaityas. These erections were either detached or cut
in the rock, and it is only the rock-cut ones of which
remains exist of an earlier date than the commencement of
the Christian era. The earliest specimen of a rock-cut
chaitya is in the Nigope cave, near Behar, constructed
about 200 B.C. This consists of two compartments, an
outer rectangular one 32 ft. 9 in. by 19 ft. 1 in., and
an inner circular one 19 ft. in diameter. The Lomas
Rishi cave is of a slightly later date: both of these
rock-cut temples exhibit in every detail a reproduction
of wooden forms. In the doorway the stone piers slope
inwards, just like raking wooden struts, and the upper
part represents the ends of longitudinal rafters supporting a roof. Later on the builders emancipated themselves to a certain extent from this servile adhesion to
older forms, and Fig. 40 gives a plan and section of a
later chaitya at Karli, near Poona. This bears a striking
resemblance to a Christian basilica: * there is first the forecourt; then a rectangular space divided by columns into
nave and aisles, and terminated by a semicircular apse.

* See Chap. X. for an illustration of a Christian Basilica.

F 2

The nave is 25 ft. 7 in. wide, and the aisles 10 ft. each; the total length is 126 ft. Fifteen columns separate the nave from the aisles, and these have bases, octagonal shafts, and rich capitals. Round the apse the columns are replaced by piers. The side aisles have flat roofs, and the central nave a stilted semicircular one, practically a vault, which

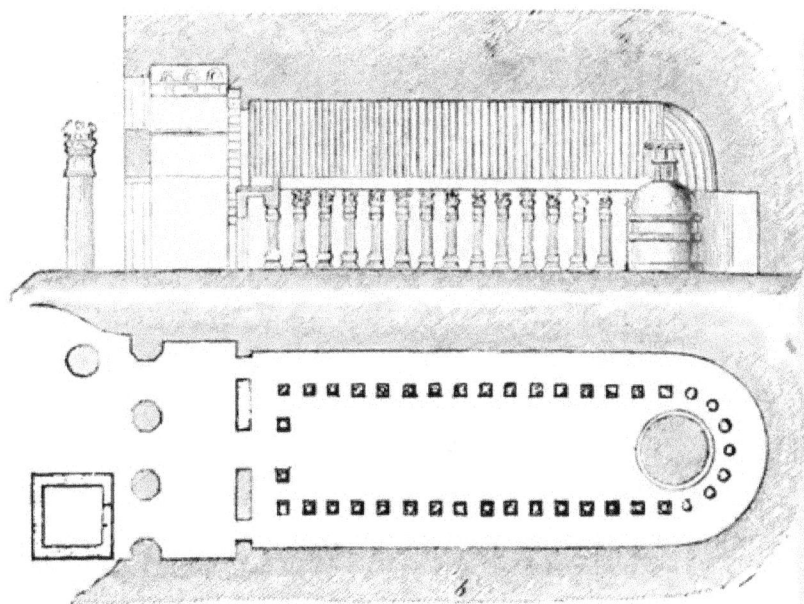

FIG. 40.—CHAITYA NEAR POONA.

at the apse becomes a semicircular dome, under which is the dagoba, the symbol of Buddhism. The screen separating the forecourt from the temple itself is richly ornamented with sculpture.

The older viharas or monasteries were also cut in the rock (Figs. 41, 42), and were divided into cells or chambers;

they were several storeys in height, and it is probable that the cells were used by devout Buddhists as habitations for the purposes of meditation.

Among the most remarkable, and in fact almost unique

FIG. 41.—THE KYLAS AT ELLORA. A ROCK-CUT MONUMENT.

features of Hindu Architecture are the so-called rails which form enclosures sometimes round the topes and sometimes round sacred trees. Occasionally they are found standing alone, though when this is the case it is

probably on account of the object which was the cause of
their erection having perished. They are built of stone,
carved so as to represent a succession of perpendicular and
horizontal bands or rails, separated by a sort of pierced
panels. The carving is of the most elaborate description,
both human and animal forms being depicted with great

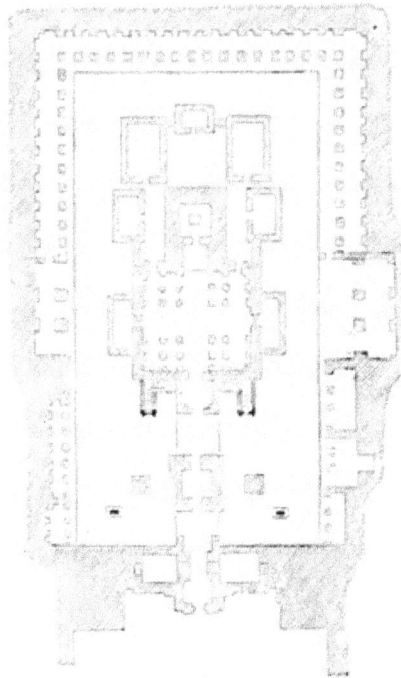

FIG. 42 —PLAN OF THE KYLAS AT ELLORA. A ROCK-CUT MONUMENT.

fidelity, and representations occur of various forms of tree
worship which have been of the greatest use in elucidating
the history of this phase of religious belief. Occasionally
the junctions of the rails are carved into a series of discs,
separated by elaborate scroll-work. These rails are fre-

quently of very large dimensions, that at Bharhut—which is one of the most recently discovered—measuring 275 ft. in circumference, with a height of 22 ft. 6 in. The date of these erections is frequently very difficult to determine, but the chief authorities gen rally concur in the opinion

FIG. 43.—VIMANA FROM MANASARA.

that none are found dating earlier than about 250 B.C., nor later than 500 A.D., so that it is pretty certain they must have been appropriated to some form of Buddhist worship.

All the buildings that we have mentioned were devoted

to the worship of Buddha, but the Jain schism, Brahmanism,
and other cults had their representative temples and buildings, a full description of which would require a volume
many times larger than the present one. Many of the
late detached buildings display rich ornamentation and
elaborate workmanship. They are generally of a pyramidal
shape, several storeys in height, covered with intricately
cut mouldings and other fantastic embellishments.

Columns are of all shapes and sizes, brackets frequently
take the place of capitals, and where capitals exist almost
every variety of fantastic form is found. It has been
stated that no fixed laws govern the plan or details of
Indian buildings, but there exists an essay on Indian
Architecture by Ram Raz—himself a Hindoo—which tends
to show that such a statement is erroneous, as he quotes
original works of considerable antiquity which lay down
stringent rules as to the planning of buildings, their height,
and the details of the columns. It is probable that a more
extended acquaintance with Hindu literature will throw
further light on these rules.

Of the various invasions which have occurred some have
left traces in the architecture of India. None of these are
more interesting than certain semi-Greek forms which are
met with in the Northern Provinces, and which without
doubt are referable to the influence of the invasion under
Alexander the Great. A far more conspicuous and widespread series of changes followed in the wake of the
Mohammedan invasions. We shall have an opportunity
later on of recurring to this subject,* but it is one to which
attention should be called at this early stage, lest it should
be thought that a large and splendid part of Indian architecture had been overlooked.

* See chapter on Saracenic Architecture.

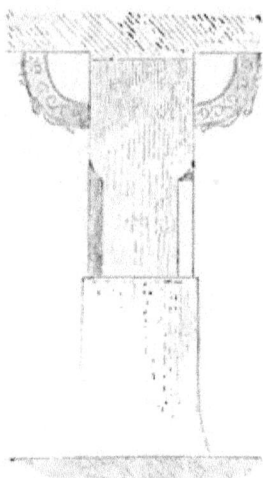

FIG. 44.—BRACKET CAPITAL.

FIG. 45.—COLUMN FROM AJUNTA.

FIG. 46.—COLUMN FROM ELLORA.

FIG. 47.—COLUMN FROM
AJUNTA.

Although the Chinese have existed as a nation continuously for between two and three thousand years, if not longer, and at a very early period had arrived at a high state of artistic and scientific cultivation, yet none of their buildings with which we are acquainted has any claim on our attention because of its antiquity. Several reasons may be assigned for this, the principal being that the Chinese seem to be as a race singularly unsusceptible to all emotions. Although they reverence their dead ancestors, yet this reverence never led them, as did that of the Egyptians, Etruscans, and other nations, to a lavish expenditure of labour or materials, to render their tombs almost as enduring as the everlasting hills. Though waves of religious zeal must have flowed over the country when Confucius inculcated his simple and practical morality and gained an influential following, and again when Buddhism was introduced and speedily became the religion of the greater portion of the people, their religious emotion never led them, as it did the Greeks and the Mediæval builders, to erect grand and lasting monuments of sacred art. When most of the Western nations were still barbarians, the Chinese had attained a settled system of government, and were acquainted with numerous scientific truths which we have prided ourselves on rediscovering within the last two centuries; but no thought ever seems to have occurred to them, as it did to the Romans, of commemorating any event connected with their life as a nation, or of handing down to posterity a record of their great achievements. Peaceful and prosperous, they have pursued the even tenor of their way at a high level of civilisation certainly, but at a most monotonous one.

The Buddhist temples of China have a strong affinity to those of India. The largest is that at Honan, the southern suburb of Canton. This is 306 ft. long by 174 ft. wide, and consists of a series of courts surrounded by colonnades and cells for the *bonzes* or priests. In the centre of the courtyard is a series of pavilions or temples connected by passages, and devoted to the worship of the idols contained in them. On each side of the main court, against the outer wall, is another court, with buildings round it, consisting of kitchen and refectories on the one side, and hospital wards on the other. It is almost certain that this is a reproduction of the earlier forms of chaityas and viharas which existed in India, and have been already referred to. The temple of Honan is two storeys in height, the building itself being of stone, but the colonnade surrounding it is of wood on marble bases. On the second storey the columns are placed on two sides only, and not all round. The columns have no capitals, but have projecting brackets. The roof of each storey projects over the columns, and has a curved section, which is, in fact, peculiar to Chinese roofs, and it is enriched at the corners with carved beasts and foliage. This is a very common form of temple throughout China.

The Taas or Pagodas are the buildings of China best known to Europeans. These are nearly always octagonal in plan, and consist generally of nine storeys, diminishing both in height and breadth as they approach the top. Each storey has a cornice composed of a fillet and large hollow moulding, supporting a roof which is turned up at every corner and ornamented with leaves and bells. On the top of all is a long pole, forming a sort of spire, surrounded by iron hoops, and supported by eight chains attached to the summit and to each angle of the roof of the topmost storey. The best known pagoda is that of

Nankin, which is 40 ft. in diameter at its base, and is faced
inside and outside with white glazed porcelain slabs keyed
into the brick core.   The roof tiles are also of porcelain,
in bands of green and yellow, and at each angle is a moulding of larger tiles, red and green alternately.   The effect
of the whole is wonderfully brilliant and dazzling.   Apart
from the coloured porcelain, nearly every portion of a
Chinese temple or pagoda is painted, colour forming the

FIG. 48.—A SMALL PAGODA.

chief means of producing effect ; but as nearly everything
is constructed of wood, there was and is no durability in
these edifices.

    In public works of utility, such as roads, canals—
one of which is nearly 700 miles in length—and boldly
designed bridges, the Chinese seem to have shown a
more enlightened mind ; and the Great Wall, which was

built to protect the northern boundary of the kingdom, about 200 B.C., is a wonderful example of engineering skill. This wall, which varies from 15 to 30 ft. in height, is about 25 ft. thick at the base, and slopes off to 20 ft. at the top. It is defended by bastions placed at stated intervals, which are 40 ft. square at the base, and about the same in height; the wall is carried altogether through a course of about 1400 miles, following all the sinuosities of the ground over which it passes. It is a most remarkable fact that a nation should have existed 2000 years ago capable of originating and completing so great a work; but it is still more remarkable that such a nation, possessing moreover, as it does, a great faculty in decorative art applied to small articles of use and fancy, should be still leading a populous and prosperous existence, and yet should have so little to show in the way of architecture, properly so termed, at the present time.

Japan, like China, possesses an architecture, but one exclusively of wood; for although the use of stone for bridges, walls, &c., had been general, all houses and temples were invariably built of wood until the recent employment of foreigners led to the erection of brick and stone buildings. The consequence has been that nearly all the old temples have been burnt down and rebuilt several times; and though it is probable that the older forms were adhered to when the buildings were re-erected, it is only by inference that we can form an idea of the ancient architecture of the country. The heavy curved roofs which are so characteristic of Chinese buildings are found also in Japan, but only in the Buddhist temples, and this makes it probable that this form of roof is not of native origin, but was introduced with the Buddhist cult. The earlier Shinto temples have a different form of roof, which is without the upward curve, but which has nearly as much projection at the eaves as the curved roofs. Where the buildings are more than one storey in height the upper is always set somewhat back, as we saw was the case in the Chinese pagodas, and considerable and pleasing variety is obtained by treating the two storeys differently. Very great skill in carving is shown, all the posts, brackets, beams, and projecting rafters being formed into elaborate representations of animals and plants, or quaintly conceived grotesques; and the flat surfaces have frequently a shallow incised arabesque pattern intertwined with foliage. The roofs are always covered with tiles, and a curious effect is produced by enriching the hips and ridges with several courses of tiles in cement, thus making them rise considerably above the other portions of the roof. A peculiar feature of Japanese houses is that the walls, whether external or internal, are not filled in with plaster, but are constructed of movable screens which slide in grooves formed in the framing of the partitions. Thus all the rooms can easily be thrown together or laid open to the outer air in hot weather. All travellers in Japan remark upon the impossibility of obtaining privacy in the hotels in consequence of this.

The Shinto temples are approached through what might be termed an archway, only that the arch does not enter into its composition. This erection is called a Torii, and is thus described by Professor Conder : *—" It is composed of two upright posts of great thickness, each consisting of the whole trunk of a tree rounded, about 15 ft. high, and placed 12 ft. apart. Across the top of these is placed a wooden lintel, projecting considerably and curving upwards at the ends. Some few feet below this another horizontal piece is tenoned into the uprights, having a

* Paper communicated to the Royal Institute of Architects.

little post in the centre helping to support the upper lintel." These erections occasionally occur in front of a Buddhist temple, when they are built of stone, exactly imitating, however, the wooden originals. This is interesting, as offering another proof, were one needed, that the curious forms of masonry exhibited in much of the work of the early nations, some of which has been described, is the result of an imitation of earlier wooden forms.

The chief effect in the buildings of the Japanese is intended to be produced by colour, which is profusely used; and they have attained to a height of perfection in the preparation of varnishes and lacquers that has never been equalled. Their lacquer is used all over their buildings, besides forming their chief means of decorating small objects. It is, however, beginning to be questioned whether the old art of lacquering is not becoming lost by the Japanese themselves, as the modern work appears by no means equal to the old. One curious form of decoration, of which the Japanese are much enamoured, consists in forming miniature representations of country scenes and landscapes; waterfalls, bridges, &c., being reproduced on the most diminutive scale. It is much to be feared that our small stock of knowledge of ancient Japanese art will never be greatly increased, as the whole country and the people are becoming modernised and Europeanised to such an extent that it appears probable there will soon be little indigenous art left in the country.

It has not been thought necessary to append to this chapter analyses of the Eastern styles similar to those which are given in the case of the great divisions of Western Architecture. The notice of these styles must unavoidably be condensed into very small space.

FIG. 40.—GREEK HONEYSUCKLE ORNAMENT.

CHAPTER V.

GREEK ARCHITECTURE.

Buildings of the Doric Order.

THE architecture of Greece has a value far higher than that attaching to any of the styles which preceded it, on account of the beauty of the buildings and the astonishing refinement which the best of them display. This architecture has a further claim on our attention, as being virtually the parent of that of all the nations of Western Europe. We cannot put a finger upon any features of Egyptian, Assyrian, or Persian architecture, the influence of which has survived to the present day, except such as were adopted by the Greeks. On the other hand, there is no feature, no ornament, nor even any principle of design which the Greek architects employed, that can be said to have now become obsolete. Not only do we find direct reproductions of Greek architecture forming part of the practice of every European country, but we are able to trace to Greek art the parentage of many of the forms and features of Roman, Byzantine, and Gothic architecture,

especially those connected with the column and which grew
out of its artistic use. Greek architecture did not include
the arch and all the forms allied to it, such as the vault
and the dome; and, so far as we know, the Greeks ab-
stained from the use of the tower. Examples of both
these features were, it is almost certain, as fully within
the knowledge of the Greeks as were those features of
Egyptian, Assyrian, and Persian buildings which they
employed; consequently it is to deliberate selection that
we must attribute this exclusion. Within the limits by
which they confined themselves, the Greeks worked with
such power, learning, taste, and skill that we may fairly
claim for their highest achievement—the Parthenon—
that it advanced as near to absolute perfection as any
work of art ever has been or ever can be carried.

Greek architecture seems to have begun to emerge from
the stage of archaic simplicity about the beginning of the
sixth century before the Christian era (600 B.C. is the
reputed date of the old Doric Temple at Corinth). All
the finest examples were erected between that date and
the death of Alexander the Great (333 B.C.), after which
period it declined and ultimately gave place to Roman.

The domestic and palatial buildings of the Greeks have
decayed or been destroyed, leaving but few vestiges. We
know their architecture exclusively from ruins of public
buildings, and to a limited extent of sepulchral monuments
remaining in Greece and in Greek colonies. By far the
most numerous and excellent among these buildings are
temples. The Greek idea of a temple was different from
that entertained by the Egyptians. The building was to
a much greater extent designed for external effect than
internal. A comparatively small sacred cell was provided
for the reception of the image of the divinity, usually with

G

one other cell behind it, which seems to have served as
treasury or sacristy; but there were no surrounding
chambers, gloomy halls, or enclosed courtyards, like those
of the Egyptian temples, visible only to persons admitted
within a jealously guarded outer wall. The temple, it is
true, often stood within some sort of precinct, but it was
accessible to all. It stood open to the sun and air; it
invited the admiration of the passer-by; its most telling
features and best sculpture were on the exterior. Whether
this may have been, to some extent, the case with Persian
buildings, we have few means of knowing, but certainly
the attention paid by the Greeks to the outside of their
temples offers a striking contrast to the practice of the
Egyptians, and to what we know of that of the Assyrians.

Fig. 50.—PLAN OF A SMALL GREEK TEMPLE IN ANTIS.

The temple, however grand, was always of simple form,
with a gable at each end, and in this respect differed
entirely from the series of halls, courts, and chambers of
which a great Egyptian temple consisted. In the very
smallest temple at least one of the gables was made into a
portico by the help of columns and two pilasters (Fig. 50).
More important temples had a larger number of columns,

and often a portico at each end (Figs. 50A and 55). The most important had columns on the flanks as well as at the front and rear, the sacred cell being, in fact, surrounded by them. It will be apparent form this that the column, together with the superstructure which rested upon it, must have played a very important part in Greek temple-architecture, and an inspection of any representations of Greek buildings will at once confirm the impression.

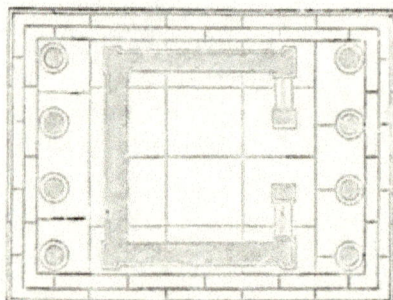

F.G. 50A.—PLAN OF A SMALL GREEK TEMPLE.

We find in Greece three distinct manners, distinguished largely by the mode in which the column is dealt with. These it would be quite consistent to call "styles," were it not that another name has been so thoroughly appropriated to them, that they would hardly now be recognised were they to be spoken of as anything else than "orders." The Greek orders are named the Doric, Ionic, and Corinthian. Each of them presents a different series of proportions, mouldings, features, and ornaments, though the main forms of the buildings are the same in all. The column and its entablature (the technical name for the frieze, architrave, and cornice, forming the usual super-structure) being the most prominent features in every

such building, have come to be regarded as the index or
characteristic from an inspection of which the order and
the degree of its development can be recognised, just as
a botanist recognises plants by their flowers. By repro-
ducing the column and entablature, almost all the characteristics of either of the orders can be copied; and hence
a technical and somewhat unfortunate use of the word
"order" to signify these features only has crept in, and has
overshadowed and to a large extent displaced its wider
meaning. It is difficult in a book on architecture to
avoid employing the word "order" when we have to speak
of a column and its entablature, because it has so often
been made use of in this sense. The student must, however, always bear in mind that this is a restricted and
artificial sense of the word, and that the column belonging
to any order is always accompanied by the use throughout
the building of the appropriate proportions, ornaments,
and mouldings belonging to that order.

The origin of Greek architecture is a very interesting
subject for inquiry, but, owing to the disappearance of
almost all very early examples of the styles, it is necessarily obscure. Such information, however, as we possess,
taken together with the internal evidence afforded by the
features of the matured style, points to the influence of
Egypt, to that of Assyria and Persia, and to an early
manner of timber construction—the forms proper to
which were retained in spite of the abandonment of
timber for marble—as all contributing to the formation
of Greek architecture.

In Asia Minor a series of monuments, many of them
rock-cut, has been discovered, which throw a curious
light upon the early growth of architecture. We refer
to tombs found in Lycia, and attributed to about the

seventh century B.C. In these we obviously have the
first work in stone of a nation of ship-builders. A Lycian
tomb—such as the one now to be seen, accurately restored,
in the British Museum—represents a structure of beams
of wood framed together, surmounted by a roof which
closely resembles a boat turned upside down. The planks,
the beams to which they were secured, and even a ridge
similar to the keel of a vessel, all reappear here, showing
that the material in use for building was so universally
timber, that when the tomb was to be "graven in the rock
for ever" the forms of a timber structure were those that
presented themselves to the imagination of the sculptor.
In other instances the resemblance to shipwrights' work
disappears, and that of a carpenter is followed by that of
the mason. Thus we find imitations of timber beams
framed together and of overhanging low-pitched roofs,
in some cases carried on unsquared rafters lying side by
side, in several of these tombs.

What happened on the Asiatic shore of the Egean must
have occurred on the Greek shores; and though none of
the very earliest specimens of reproduction in stone of
timber structures has come down to us, there are abundant
traces, as we shall presently see, of timber originals in
buildings of the Doric order. Timber originals were not,
however, the only sources from which the early inhabitants of Greece drew their inspiration.

Constructions of extreme antiquity, and free from any
appearance of imitating structures of timber, mark the sites
of the oldest cities of Greece, Mycenæ and Orchomenos for
example, the most ancient being Pelasgic city walls of unwrought stone (Fig. 51). The so-called Treasury of Atreus
at Mycenæ, a circular underground chamber 48 ft. 6 in.
in diameter, and with a pointed vault, is a well-known
specimen of more regular yet archaic building. Its vault

is constructed of stones corbelling over one another, and
is not a true arch (Figs. 52, 52A). The treatment of an
ornamental column found here, and of the remains of

FIG. 51.—ANCIENT GREEK WALL OF UNWROUGHT STONE FROM SAMOTHRACE.

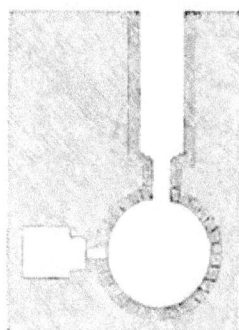

FIG. 52—PLAN OF THE TREASURY
OF ATREUS AT MYCENÆ.

FIG. 52A.— SECTION OF THE TREASURY OF
ATREUS AT MYCENÆ.

sculptured ornaments over a neighbouring gateway called
the Gate of the Lions, is of very Asiatic character, and
seems to show that whatever influences had been brought
to bear on their design were Oriental.

A wide interval of time and a great contrast in taste separate the early works of Pelasgic masonry and even the chamber at Mycenæ from even the rudest and most archaic of the remaining Hellenic works of Greece. The

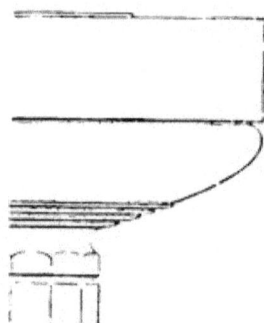

FIG. 53.—GREEK DORIC CAPITAL FROM SELINUS.

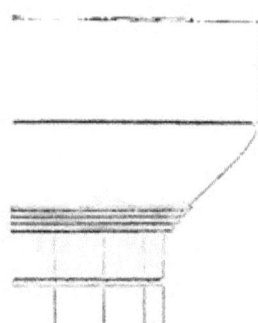

FIG. 53A.—GREEK DORIC CAPITAL FROM THE THESEUM.

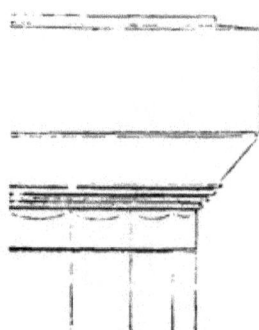

FIG. 53B.—GREEK DORIC CAPITAL FROM SAMOTHRACE.

Doric temple at Corinth is attributed, as has been stated, to the seventh century B.C. This was a massive masonry structure with extremely short, stumpy columns, and strong mouldings, but presenting the main features of the Doric style, as we know it, in its earliest and rudest form.

Successive examples (Figs. 53 to 53B) show increasing slenderness of proportions and refinement of treatment, and are accompanied by sculpture which approaches nearer and nearer to perfection; but in the later and best buildings, as in the earliest and rudest, certain forms are retained for which it seems impossible to account, except on the supposition that they are reproductions in stone or marble of a timber construction. These occur in the entablature, while the column is of a type which it is hard to believe is not copied from originals in use in Egypt many centuries earlier, and already described (chap. ii.).

We will now proceed to examine a fully-developed Greek Doric temple of the best period, and in doing so we shall be able to recognise the forms referred to in the preceding paragraph as we come to them. The most complete Greek Doric temple was the Parthenon, the work of the architect Ictinus, the temple of the Virgin Goddess Athene (Minerva) at Athens, and on many accounts this building will be the best to select for our purpose.*

The Parthenon at Athens stood on the summit of a lofty rock, and within an irregularly shaped enclosure, something like a cathedral close; entered through a noble gateway.†
The temple itself was of perfectly regular plan, and stood quite free from dependencies of any sort. It consisted of a cella, or sacred cell, in which stood the statue of the goddess, with one chamber (the treasury) behind. In the cella, and also in the chamber behind, there were columns. A series of columns surrounded this building, and at either end was a portico, eight columns wide, and two deep. There were two pediments, or gables, of flat pitch, one at each end. The whole stood on a basement

---

* See Frontispiece and Figs. 54 and 55.　　　† The Propylæa.

Fig. 51.—Ruins of the Parthenon at Athens.

of steps; the building, exclusive of the steps, being
228 ft. long by 101 ft. wide, and 64 ft. high. The columns

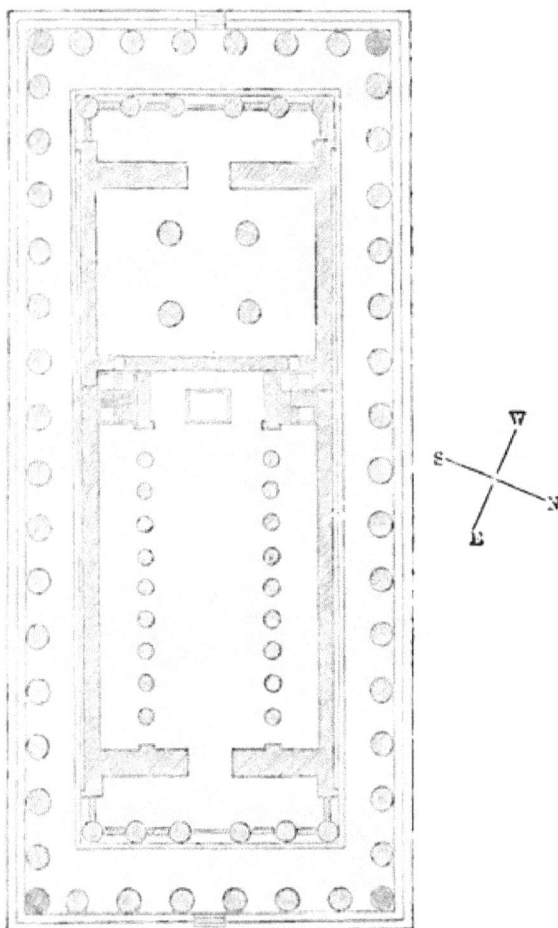

FIG. 55.—PLAN OF THE PARTHENON AT ATHENS.

were each 34 ft. 3 in. high, and more than 6 ft. in diameter
at the base; a portion of the shaft and of the capital of

one is in the British Museum, and a magnificent reproduction, full size, of the column and its entablature may
be seen at the École des Beaux Arts, Paris. The ornaments consisted almost exclusively of sculpture of the
very finest quality, executed by or under the superin-
tendence of Pheidias. Of this sculpture many specimens
are now in the British Museum.

FIG. 56.—THE ROOF OF A GREEK DORIC TEMPLE, SHOWING THE MARBLE TILES.

The construction of this temple was of the most solid
and durable kind, marble being the material used; and
the workmanship was most careful in every part of which
remains have come down to us. The roof was, no doubt,
made of timber and covered with marble tiles (Fig. 56),
carried on a timber framework, all traces of which have entirely perished; and the mode in which it was constructed
is a subject upon which authorities differ, especially as

to what provision was made for the admission of light. The internal columns, found in other temples as well as in the Parthenon, were no doubt employed to support this roof, as is shown in Bötticher's restoration of the Temple at Pæstum which we reproduce (Fig. 56A), though without pledging ourselves to its accuracy ; for, indeed, it

FIG. 56A.—SECTION OF THE GREEK DORIC TEMPLE AT PÆSTUM. AS RESTORED BY BÖTTICHER.

seems probable that something more or less like the clerestory of a Gothic church must have been employed to admit light to these buildings, as we know was the case in the Hypostyle Hall at Karnak. But this structure, if it existed, has entirely disappeared.*

The order of the Parthenon was Doric, and the leading proportions were as follows :—The column was 5·56

* Mr. Fergusson's investigations, soon, it is understood, to be published in a complete form, clearly show that the clerestory and roof can be restored with the greatest probability.

Architrave . .

Capital { Abacus
         { Echinus

Shaft or Column

Stylobate . . .

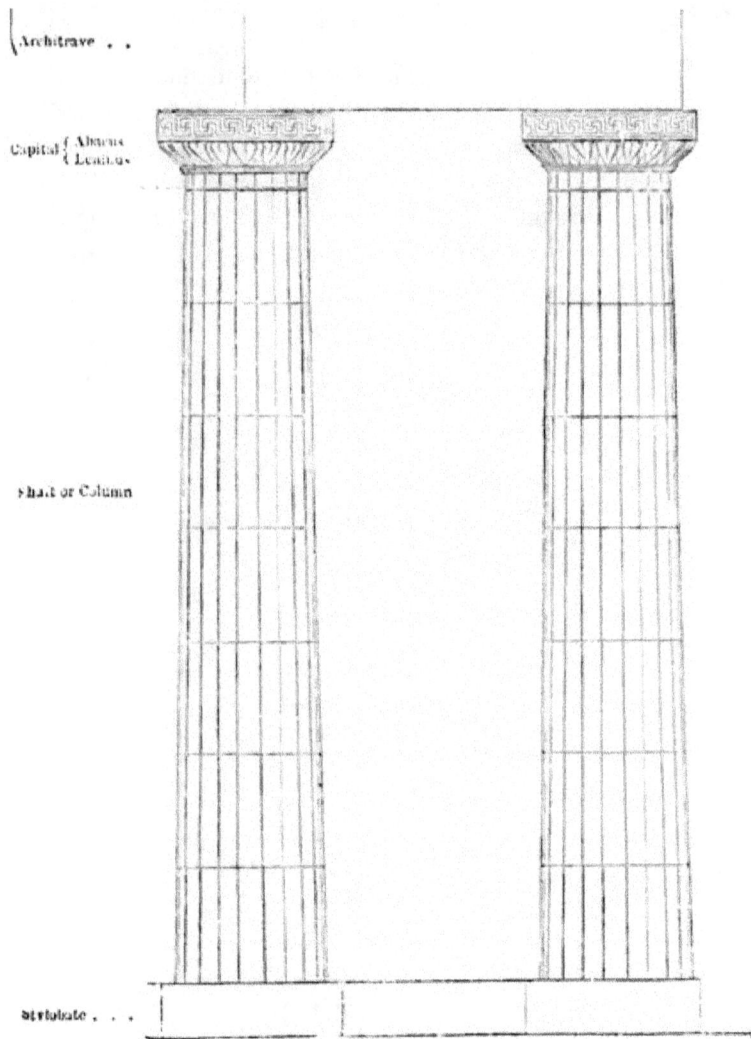

FIG. 57.—THE GREEK DORIC ORDER FROM THE THESEUM.

diameters high; the whole height, including the stylobate or steps, might be divided into nine parts, of which two go to the stylobate, six to the column, and one to the entablature.

The Greek Doric order is without a base; the shaft of the column springs from the top step and tapers towards the top, the outline being not, however, straight, but of a subtle curve, known technically as the entasis of the column. This shaft is channelled with twenty shallow channels,* the ridges separating one from another being

Fig. 59.—The Fillets under a Greek Doric Capital.

Fig. 58.—Plan of a Greek Doric Column.

very fine lines. A little below the moulding of the capital, fine sinkings, forming lines round the shaft, exist, and above these the channels of the flutes are stopped by or near the commencement of the projecting moulding of the capital. This moulding, which is of a section calculated to convey the idea of powerful support, is called the echinus, and its lower portion is encircled by a series of fillets (Fig. 59), which are cut into it. Above the echinus, which is circular, like the shaft, comes the highest member — the abacus, a square stout slab of marble, which completes the capital of the column. The

* In a few instances a smaller number is found.

whole is most skilfully designed to convey the idea of
sturdy support, and yet to clothe the support with grace.
The strong proportions of the shaft, the slight curve
of its outline, the lines traced upon its surface by the
channels, and even the vigorous uncompromising planting
of it on the square step from which it springs, all contribute to made the column look strong. The check given
to the vigorous upward lines of the channels on the shaft
by the first sinkings, and their arrest at the point where
the capital spreads out, intensified as it is by the series of

FIG. 60.—CAPITAL OF A GREEK DORIC COLUMN FROM ÆGINA,
WITH COLOURED DECORATION.

horizontal lines drawn round the echinus by the fillets
cut into it, all seem to convey the idea of spreading the
supporting energy of the column outwards; and the abacus
appears naturally fitted, itself inert, to receive a burden
placed upon it and to transmit its pressure to the capital
and shaft below.

The entablature which formed the superstructure consisted first of a square marble beam—the architrave,
which, it may be assumed, represents a square timber

beam that occupied the same position in the primitive

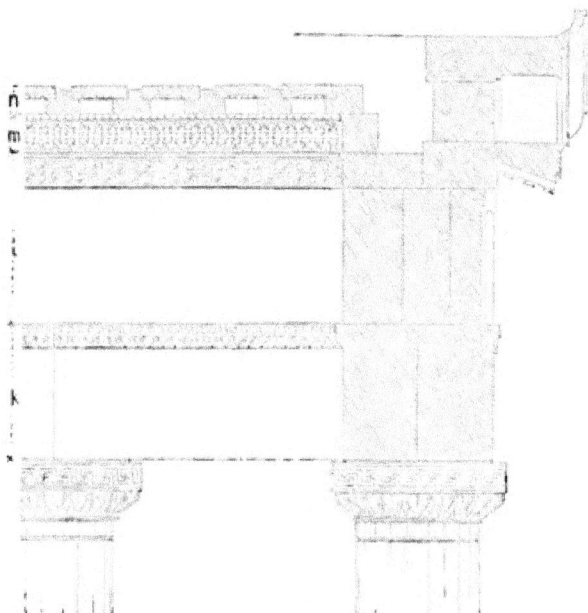

FIG. 61.—SECTION OF THE ENTABLATURE OF THE GREEK DORIC ORDER

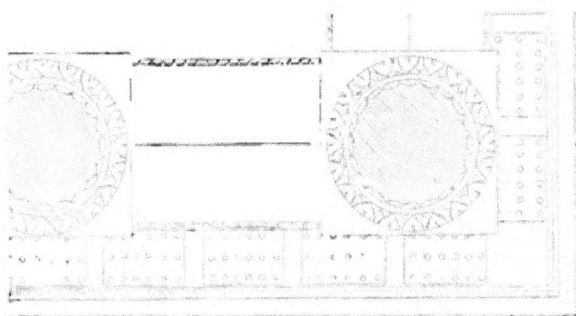

FIG. 62.—PLAN, LOOKING UP, OF PART OF A GREEK DORIC PERISTYLE.

structures. On this rests a second member called the frieze, the prominent feature of which is a series of slightly projecting features, known as triglyphs (three channels) (Fig. 63), from the channels running down their face. These closely resemble, and no doubt actually represent, the ends of massive timber beams, which must have connected the colonnade to the wall of the cell in earlier buildings. At the bottom of each is a row of small pendants, known as guttæ, which closely resemble wooden pins, such as would be used to keep a timber beam in place. The panels between the triglyphs are

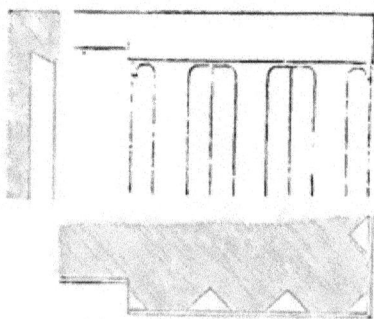

FIG. 63.—DETAILS OF THE TRIGLYPH.

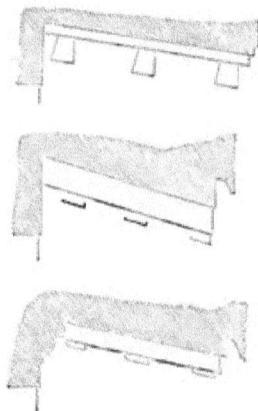

FIG. 64.—DETAILS OF THE MUTULES.

usually as wide as they are high. They are termed metopes, and sculpture commonly occupies them. The third division of the entablature, the cornice, represents the overhanging eaves of the roof.

The cornices employed in classic architecture may be almost invariably subdivided into three parts : the supporting part, which is the lowest,—the projecting part, which is the middle,—and the crowning part, which is the highest division of the cornice. The supporting part in a Greek Doric cornice is extremely small. There are no mouldings,

H

such as we shall find in almost every other cornice, calculated to convey the idea of contributing to sustain the projection of the cornice, but there are slabs of marble, called mutules (Fig. 64), dropping towards the outer end, of which one is placed over each triglyph and one between every two. These seem to recall, by their shape, their position, and their slope alike, the ends of the rafters of a timber roof; and their surface is covered with small projections which resemble the heads of wooden pins, similar to those already alluded to. The projecting part, in this as in almost all cornices, is a plain upright face of some height, called "the corona," and recalling probably a "facia" or flat narrow board such as a carpenter of the present day would use in a similar position, secured in the original structure to the ends of the rafters and supporting the eaves. Lastly, the crowning part is, in the Greek Doric, a single convex moulding, not very dis-imilar in profile to the ovolo of the capital, and forming what we commonly call an eaves-gutter.

At the ends of the building the two upper divisions of the cornice—namely, the projecting corona and the crowning ovolo—are made to follow the sloping line of the gable, a second corona being also carried across horizontally in a manner which can be best understood by inspecting a diagram of the corner of a Greek Doric building (Fig. 57); and the triangular space thus formed was termed a pediment, and was the position in which the finest of the sculpture with which the building was enriched was placed.

In the Parthenon a continuous band of sculpture ran round the exterior of the cell, near the top of the wall.

One other feature was employed in Greek temple-architecture. The *anta* was a square pillar or pier of masonry attached to the wall, and corresponded very closely to our pilaster; but its capital always differed from that of the columns in the neighbourhood of which it was employed. The antæ of the Greek Doric order, as employed in the Parthenon, have a moulded base, which it will be remembered is not the case with the column, and their capital has for its principal feature an under-cut moulding, known as the bird's beak, quite dissimilar from the ovolo of the capital of the column (Fig. 65). Sometimes the portico of a temple consisted of the side walls prolonged, and ending in two antæ, with two or more columns standing between them. Such a portico is said to be in antis.

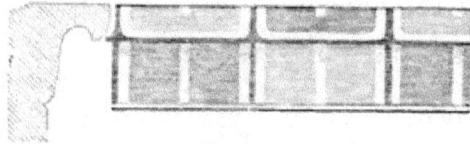

FIG. 65.—ELEVATION AND SECTION OF THE CAPITAL OF A GREEK ANTA, WITH COLOURED DECORATIONS.

The Parthenon presents examples of the most extraordinary refinements in order to correct optical illusions. The delicacy and subtlety of these are extreme, but there can be no manner of doubt that they existed. The best known correction is the diminution in diameter or taper, and the *entasis* or convex curve of the tapered outline of the shaft of the column. Without the taper, which is perceptible enough in the order of this building, and much more marked in the order of earlier buildings, the columns would look top-heavy; but the entasis is an additional optical correction to prevent their outline from appearing hollowed, which it would have done had there been no curve. The columns of the Parthenon have shafts that are over 34 ft. high, and diminish from a diameter of 6·15 ft. at

H 2

the bottom to 4·81 ft. at the top. The outline between these points is convex, but so slightly so that the curve departs at the point of greatest curvature not more than ¾ in. from the straight line joining the top and bottom. This is, however, just sufficient to correct the tendency to look hollow in the middle.

A second correction is intended to overcome the apparent tendency of a building to spread outwards towards the top. This is met by inclining the columns slightly inwards. So slight, however, is the inclination, that were the axes of two columns on opposite sides of the Parthenon continued upwards till they met, the meeting-point would be 1952 yards, or, in other words, more than one mile from the ground.

Another optical correction is applied to the horizontal lines. In order to overcome a tendency which exists in all long lines to seem as though they droop in the middle, the lines of the architrave, of the top step, and of other horizontal features of the buildings, are all slightly curved. The difference between the outline of the top step of the Parthenon and a straight line joining its two ends is at the greatest only just over 2 inches.

The last correction which it is necessary to name here was applied to the vertical proportions of the building. The principles upon which this correction rests have been demonstrated by Mr. John Pennethorne;[*] and it would hardly come within the scope of this volume to attempt to state them here: suffice it to say, that small additions, amounting in the entire height of the order to less than 5 inches, were made to the heights of the various members of the order, with a view to secure that from one definite point of view the effect of foreshortening

* 'Geometry and Optics of Ancient Architecture.'

should be exactly compensated, and so the building should appear to the spectator to be perfectly proportioned.

The Parthenon, like many, if not all Greek buildings, was profusely decorated with coloured ornaments, of which nearly every trace has now disappeared, but which must have contributed largely to the splendid beauty of the building as a whole, and must have emphasised and set off its parts. The ornaments known as Doric frets were largely employed. They consist of patterns made entirely of straight lines interlacing, and, while preserving the severity which is characteristic of the style, they permit of the introduction of considerable richness.

The principal remaining examples or fragments of Greek Doric may be enumerated as follows:—

In Greece.

Temple of (?) Athena, at Corinth, ab. 650 B.C.
Temple of (?) Zeus, in the island of Ægina, ab. 550 B.C.
Temple of Theseus (Theseium), at Athens, 465 B.C.
Temple of Athena (Parthenon), on the Acropolis at Athens, fin. 438 B.C.
The Propylæa, on the Acropolis at Athens, 436–431 B.C.
Temple of Zeus at Olympia.
Temple of Apollo Epicurius, at Bassæ,* in Arcadia (designed by Ictinus).
Temple of Apollo Epicurius, at Phigaleia. in Arcadia (built by Ictinus).
Temple of Athena, on the rock of Sunium, in Attica.
Temple of Nemesis, at Rhamnus, in Attica.
Temple of Demeter (Ceres), at Eleusis, in Attica.

In Sicily and South Italy.

Temple of (?) Zeus, at Agrigentum, in Sicily (begun B.C. 480).
Temple of Ægesta (or Segesta), in Sicily.
Temple of (?) Zeus, at Selinus, in Sicily (? ab. 410 B.C.).
Temple of (?) Athena, at Syracuse, in Sicily.
Temple of Poseidon, at Pæstum, in South of Italy (? ab. 550 B.C.).

---

* ? Exterior Doric—Interior Ionic.

Fig. 66.—Palmette and Honeysuckle.

CHAPTER VI.

GREEK ARCHITECTURE.

Buildings of the Ionic and Corinthian Orders.

THE Doric was the order in which the full strength and the complete refinement of the artistic character of the Greeks were most completely shown. There was a great deal of the spirit of severe dignity proper to Egyptian art in its aspect; but other nationalities contributed to the formation of the many-sided Greek nature, and we must look to some other country than Egypt for the spirit which inspired the Ionic order. This seems to have been brought into Greece by a distinct race, and shows marks of an Asiatic origin. The feature which is most distinctive is the one most distinctly Eastern—the capital of the column, ornamented always by volutes, *i.e.* scrolls, which bear a close resemblance to features similarly employed in the columns found at Persepolis. The same resemblance can be also detected in the moulded bases, and even the shafts of the columns, and in many of the ornaments employed throughout the buildings.

In form and disposition an ordinary Ionic temple was similar to one of the Doric order, but the general proportions are more slender, and the mouldings of the order are more numerous and more profusely enriched. The column in the Ionic order had a base, often elaborately

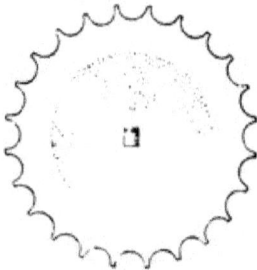

FIG. 67.—SHAFT OF IONIC COLUMN SHOWING THE FLUTES.

FIG. 69.—IONIC CAPITAL. SIDE ELEVATION.

FIG. 68.—IONIC CAPITAL. FRONT ELEVATION.

and sometimes singularly moulded (Figs. 74, 75). The shaft (Figs. 67, 70) is of more slender proportions than the Doric shaft. It was fluted, but its channels are more numerous, and are separated from one another by broader

fillets than in the Doric. The distinctive feature, as in
all the orders, is the capital (Figs. 68, 69), which is
recognised at a glance by the two remarkable ornaments
already alluded to as like scrolls, and known as volutes.
These generally formed the faces of a pair of cushion-
shaped features, which could be seen in a side view of
the capital; but sometimes volutes stand in a diagonal
position, and in almost every building they differ slightly.
The abacus is less deep than in the Greek Doric, and
it is always moulded at the edge, which was never the
case with the Doric abacus. The entablature (Fig. 70)
is, generally speaking, richer than that of the Doric order.
The architrave, for example, has three facias instead of
being plain. On the other hand, the frieze has no triglyphs,
and but rarely sculpture. There are more members in
the cornice, several mouldings being combined to fortify
the supporting portion. These have sometimes been
termed "the bed mouldings;" and among them occurs one
which is almost typical of the order, and is termed a
dentil band. This moulding presents the appearance of
a plain square band of stone, in which a series of cuts had
been made dividing it into blocks somewhat resembling
teeth, whence the name. Such an ornament is more
naturally constructed in wood than in stone or marble, but
if the real derivation of the Ionic order, as of the Doric,
be in fact from timber structures, the dentil band is
apparently the only feature in which that origin can now
be traced. The crowning member of the cornice is a
partly hollow moulding, technically called a "cyma recta,"
less vigorous than the convex ovolo, of the Doric: this
moulding, and some of the bed mouldings, were commonly enriched with carving. Altogether more slenderness and less vigour, more carved enrichment and less
painted decoration, more reliance on architectural orna-

FIG. 70.—THE IONIC ORDER. FROM PRIENE, ASIA MINOR.

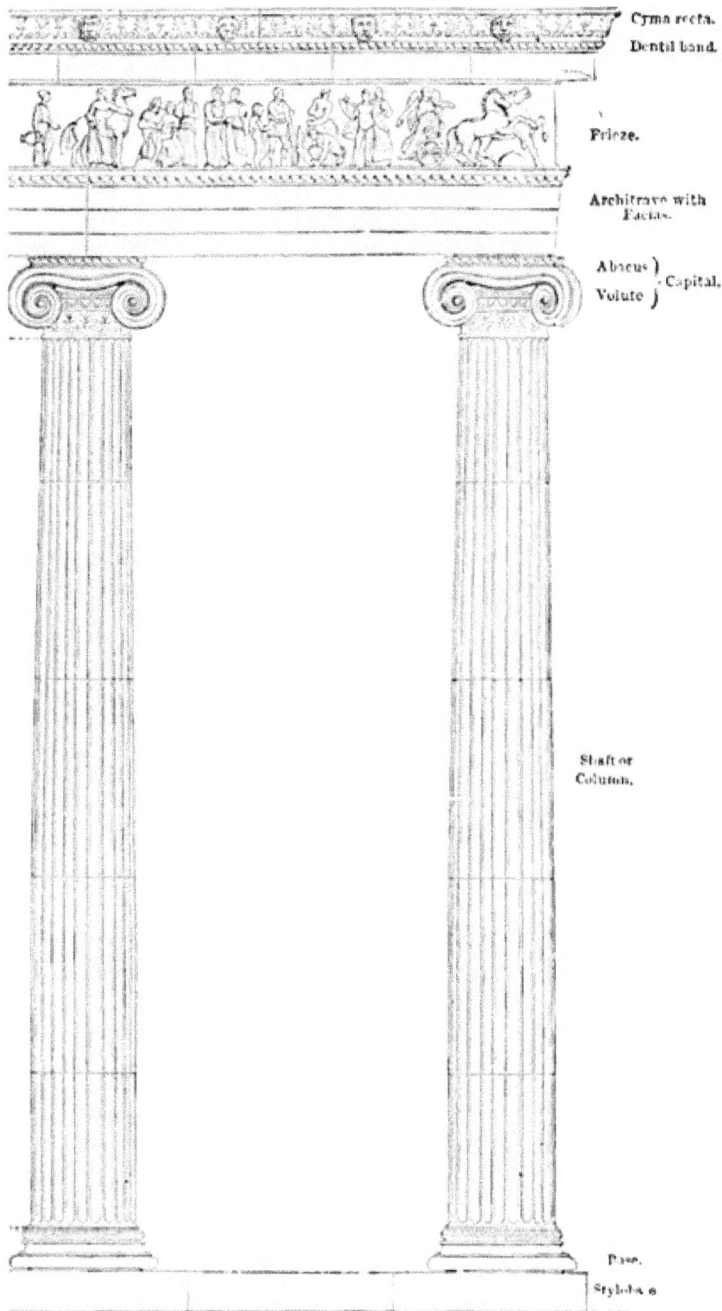

Cyma recta.

Dentil band.

Frieze.

Architrave with Facias.

Abacus
Volute } Capital.

Shaft or Column.

Base.

Stylobate.

FIG 71.—IONIC ORDER. FROM THE ERECHTHEIUM, ATHENS.

ment and less on the work of the sculptor, appear to distinguish those examples of Greek Ionic which have come down to us, as compared with Doric buildings.

The most numerous examples of the Ionic order of which remains exist are found in Asia Minor, but the most refined and complete is the Erechtheium at Athens

FIG. 72.—NORTH-WEST VIEW OF THE ERECHTHEIUM, IN THE TIME OF PERICLES.

(Figs. 72, 73), a composite structure containing three temples built in juxtaposition, but differing from one another in scale, levels, dimensions, and treatment. The principal order from the Erechtheium (Fig. 71) shows a large amount of enrichment introduced with the most refined and severe taste. Specially remarkable are the ornaments (borrowed from the Assyrian honeysuckle)

which encircle the upper part of the shaft at the point
where it passes into the capital, and the splendid spirals
of the volutes (Figs. 68, 69). The basis of the columns

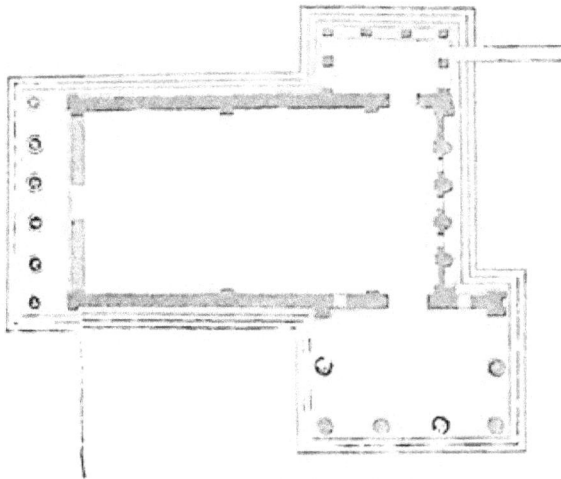

FIG. 73.—PLAN OF THE ERECHTHEIUM.

FIG. 74.—IONIC BASE FROM THE
TEMPLE OF THE WINGLESS VIC-
TORY (NIKÉ APTEROS).

FIG. 75.—IONIC BASE MOULDINGS
FROM PRIENE.

in the Erechtheium example are models of elegance and
beauty. Those of some of the examples from Asia Minor

are overloaded with a vast number of mouldings, by no means always producing a pleasing effect (Figs. 74, 75). Some of them bear a close resemblance to the bases of the columns at Persepolis.

The most famous Greek building which was erected in the Ionic style was the Temple of Diana at Ephesus. This temple has been all but totally destroyed, and the very site of it had been for centuries lost and unknown till the energy and sagacity of an English architect (Mr. Wood) enabled him to discover and dig out the vestiges of the building. Fortunately sufficient traces of the foundation have remained to render it possible to recover the plan of the temple completely; and the discovery of fragments of the order, together with representations on ancient coins and a description by Pliny, have rendered it possible to make a restoration on paper of the general appearance of this famous temple, which must be very nearly, if not absolutely, correct.

The walls of this temple enclosed, as usual, a cella (in which was the statue of the goddess), with apparently a treasury behind it: they were entirely surrounded by a double series of columns, with a pediment at each end. The exterior of the building, including these columns, was about twice the width of the cella. The whole structure, which was of marble, was planted on a spacious platform with steps. The account of Pliny refers to thirty-six columns, which he describes as "*columnæ celatæ*" (sculptured columns), adding that one was by Scopas, a very celebrated artist. The fortunate discovery by Mr. Wood of a few fragments of these columns shows that the lower part of the shaft immediately above the base was enriched by a group of figures — about life-size — carved in the boldest relief and encircling the column. One of these groups has been brought to the British Museum, and its

beauty and vigour enable the imagination partly to restore
this splendid feature, which certainly was one of the most
sumptuous modes of decorating a building by the aid of
sculpture which has ever been attempted; and the effect
must have been rich beyond description.

It is worth remark that the Erechtheium, which has
been already referred to, contains an example of a different,
and perhaps a not less remarkable, mode of combining
sculpture with architecture. In one of its three porticoes
(Fig. 72) the columns are replaced by standing female
figures, known as caryatidæ, and the entablature rests on
their heads. This device has frequently been repeated in
ancient and in modern architecture, but, except in some
comparatively obscure examples, the sculptured columns
of Ephesus do not appear to have been imitated.

Another famous Greek work of art, the remains of
which have been, like the Temple of Diana, disinterred by
the energy and skill of a learned Englishman, belonged
to the Ionic order. To Mr. Newton we owe the recovery
of the site, and considerable fragments of the architectural
features, of the Mausoleum of Halicarnassus, one of the
ancient wonders of the world. The general outline of
this monument must have resembled other Greek tombs
which have been preserved, such, for example, as the Lion
Tomb at Cnidus; that is to say, the plan was square:
there was a basement, above this an order, and above that
a steep pyramidal roof rising in steps, not carried to a
point, but stopping short to form a platform, on which
was placed a quadriga (or four-horsed chariot). This
building is known to have been richly sculptured, and
many fragments of great beauty have been recovered.
Indeed it was probably its elaboration, as well as its
very unusual height (for the Greek buildings were seldom
lofty), which led to its being so celebrated.

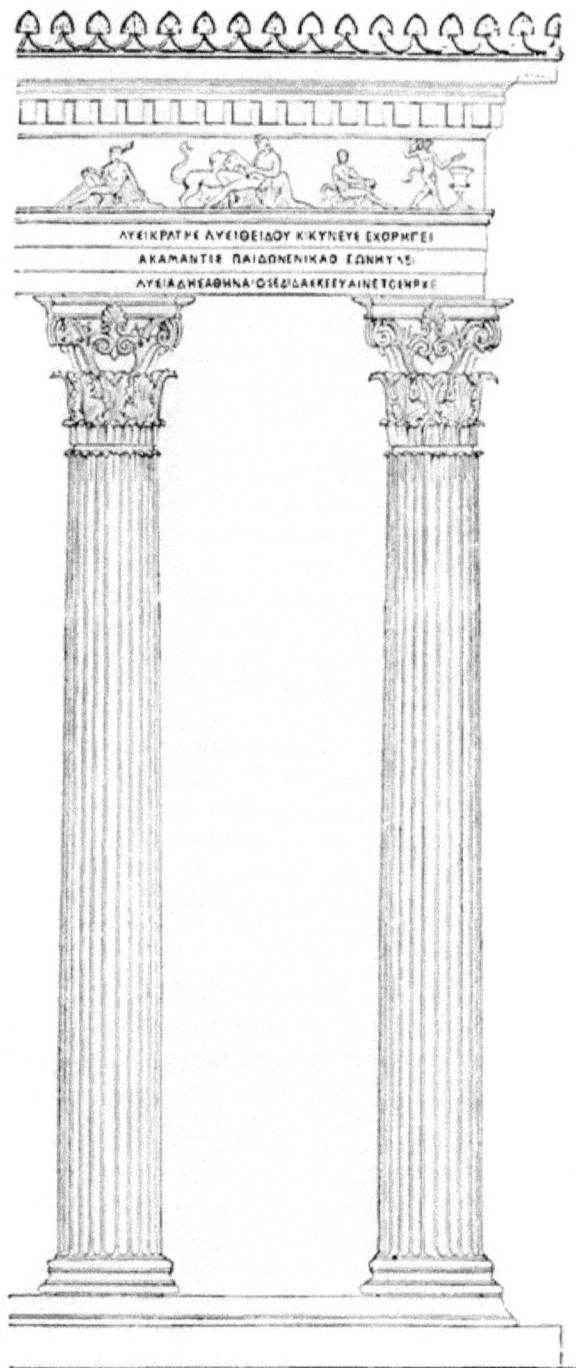

ΛΥΣΙΚΡΑΤΗΣ ΛΥΣΙΘΕΙΔΟΥ ΚΙΚΥΝΕΥΣ ΕΧΟΡΗΓΕΙ
ΑΚΑΜΑΝΤΙΣ ΠΑΙΔΩΝΕΝΙΚΑΟ ΕΩΝΗΥΛΕΙ
ΛΥΣΙΑΔΗΣΑΘΗΝΑΙΟΣΕΔΙΔΑΣΚΕΕΥΑΙΝΕΤΟΣΗΡΧΕ

FIG. 76.—THE CORINTHIAN ORDER. FROM THE MONUMENT OF LYSICRATES AT ATHENS.

The Corinthian order, the last to make its appearance, was almost as much Roman as Greek, and is hardly found in any of the great temples of the best period of which remains exist in Greece, though we hear of its use. For example, Pausanias states that the Corinthian order was employed in the interior of the Temple of Athena Alea at Tegea, built by Scopas, to which a date shortly after the year 394 B.C. is assigned. The examples which we possess

Fig. 77.—Corinthian Capital from the Monument of Lysicrates at Athens.

are comparatively small works, and in them the order resembles the Ionic, but with the important exceptions that the capital of the column is quite different, that the proportions are altogether a little slenderer, and that the enrichments are somewhat more florid.

The capital of the Greek Corinthian order, as seen in the Choragic Monument of Lysicrates at Athens (Fig. 78) —a comparatively miniature example, but the most perfect we have—is a work of art of marvellous beauty (Fig. 77).

FIG. 76.—MONUMENT TO LYSICRATES AT ATHENS, AS IN THE TIME OF PERICLES.

1

It retains a feature resembling the Ionic volute, but reduced to a very small size, set obliquely and appearing to spring from the sides of a kind of long bell-shaped termination to the column. This bell is clothed with foliage, symmetrically arranged and much of it studied, but in a conventional manner, from the graceful foliage of the acanthus; between the two small volutes appears an Assyrian honeysuckle, and tendrils of honeysuckle, conventionally treated, occupy part of the upper portion of the capital. The abacus is moulded, and is curved on plan, and the base of the capital is marked by a very unusual turning-down of the flutes of the columns. The entire structure to which this belonged is a model of elegance, and the large sculptured mass of leaves and tendrils with which it is crowned is especially noteworthy.

FIG. 79.—CAPITAL OF ANTÆ FROM MILETUS. SIDE VIEW.

A somewhat simpler Corinthian capital, and another of very rich design, are found in the Temple of Apollo Didymaeus at Miletus, where also a very elegant capital for the antae—or pilasters—is employed (Figs. 79, 81). A more ornamental design for a capital could hardly be adopted than that of the Lysicrates example, but there was room for more elaboration in the entablature, and accordingly large

FIG. 80.—RESTORATION OF THE GREEK THEATRE OF SEGESTA.

richly-sculptured brackets seem to have been introduced, and a profusion of ornament was employed. The examples of this treatment which remain are, however, of Roman origin rather than Greek.

The Greek cities must have included structures of great beauty and adapted to many purposes, of which in most cases few traces, if any, have been preserved. We have no remains of a Greek palace, or of Greek dwelling-houses, although those at Pompeii were probably erected and decorated by Greek artificers, for Roman occupation. The agora of a Greek city, which was a place of public assembly something like the Roman Forum, is known to us only by descriptions in ancient writers, but we possess some remains of Greek theatres; and from these, aided by Roman examples and written descriptions, can understand what these buildings were. The auditory was curved in plan, occupying rather more than a semicircle; the seats rose in tiers one behind another; a circular space was reserved for the chorus in the centre of the seats, and behind it was a raised stage, bounded by a wall forming its back and sides : a rough notion of the arrange-ment can be obtained from the lecture theatre of many modern colleges, and our illustration (Fig. 80) gives a general idea of what must have been the appearance of one of these structures. Much of the detail of these buildings is, however, a matter of pure speculation, and consequently does not enter into the scheme of this manual.

FIG. 81.—CAPITAL OF ANTÆ FROM MILETUS.

CHAPTER VII.

GREEK ARCHITECTURE.

Analysis.

THE *Plan* or floor-disposition of a Greek building was always simple however great its extent, was well judged for effect, and capable of being understood at once. The grandest results were obtained by simple means, and all confusion, uncertainty, or complication were scrupulously avoided. Refined precision, order, symmetry, and exactness mark the plan as well as every part of the work.

The plan of a Greek temple may be said to present many of the same elements as that of an Egyptian temple, but, so to speak, turned inside out. Columns are relied on by the Greek artist, as they were by the Egyptian artist, as a means of giving effect; but they are placed by him outside the building instead of within its courts and halls. The Greek, starting with a comparatively small nucleus formed by the cell and the treasury, encircles them by a magnificent girdle of pillars, and so makes a grand structure, the first hint or suggestion being in all probability to be found in certain small Egyptian buildings to which reference has already been made. The disposition of these columns and of the great range of steps, or stylobate, is the most marked feature in Greek temple plans. Columns also existed, it is true, in the interior of the building, but these were of smaller size, and seem to have been introduced to aid in carrying the roof and the clerestory, if there was one. They have in several instances disappeared, and there is certainly no ground for supposing that in any Greek interior the grand but oppressive effect of a hypostyle hall was attempted to be reproduced. That was abandoned, together with the complication, seclusion, and gloom of the long series of chambers, cells, &c., placed one behind another, just as the contrasts and surprises of the series of courts and halls following in succession were abandoned for the one simple but grand mass built to be seen from without rather than from within. In the greater number of Greek buildings a degree of precision is exhibited, to which the Egyptians did not attain. All right angles are absolutely true ; the setting-out (or spacing) of the different columns, piers, openings, &c., is perfectly exact; and, in the Parthenon, the patient investigations of Mr. Penrose and other skilled observers have disclosed a degree of accuracy as well as refinement which resembles the precision with which astronomical instruments are adjusted in Europe at the present day, rather than the rough-and-ready measurements of a modern mason or bricklayer.

What the plans of Greek palaces might have exhibited, did any remains exist, is merely matter for inference and conjecture, and it is not proposed in this volume to

pass far beyond ascertained and observed facts. There
can be, however, little doubt that the palaces of the West
Asiatic style must have at least contributed suggestions
as to internal disposition of the later and more magnificent Greek mansions. The ordinary dwelling-houses of
citizens, as described by ancient writers, resembled those
now visible in the disinterred cities of Pompeii and
Herculaneum, which will be referred to under Roman
Architecture.* The chief characteristic of the plan of
these is that they retain the disposition which in the
temples was discarded; that is to say, all the doors and
windows looked into an inner court, and the house was
as far as possible secluded within an encircling wall.
The contrast between the openness of the public life
led by the men in Greek cities, and the seclusion of the
women and the families when at home, is remarkably
illustrated by this difference between the public and
private buildings.

The plan of the triple building called the Erechtheium
(Fig. 72) deserves special mention, as an example of an
exceptional arrangement which appears to set the ordinary
laws of symmetry at defiance, and which is calculated to
produce a result into which the picturesque enters at least
as much as the beautiful. Though the central temple is
symmetrical, the two attached porticoes are not so, and
do not, in position, dimensions, or treatment, balance one
another. The result is a charming group, and we cannot
doubt that other examples of freedom of planning would
have been found, had more remains of the architecture of
the great cities of Greece come down to our own day.

In public buildings other than temples—such as the

* See Chap. IX.

theatre, the agora, and the basilica—the Greek architects seem to have had great scope for their genius; the planning of the theatres shows skilful and thoroughly complete provisions to meet the requirements of the case. A circular disposition was here introduced—not, it is true, for the first time, since it is rendered probable by the representations on sculptured slabs that some circular buildings existed in Assyria, and circular buildings remain in the archaic works at Mycenæ; but it was now elaborated with remarkable completeness, beauty, and mastery over all the difficulties involved. Could we see the great theatre of Athens as it was when perfect, we should probably find that as an interior it was almost unrivalled, alike for convenience and for beauty; and for these excellences it was mainly indebted to the elegance of its planning. The actual floor of many of the Greek temples appears to have been of marble of different colours.

The Walls.

The construction of the walls of the Greek temples rivalled that of the Egyptians in accuracy and beauty of workmanship, and resembled them in the use of solid materials. The Greeks had within reach quarries of marble, the most beautiful material which nature has provided for the use of the builder; and great fineness of surface and high finish were attained. Some interesting examples of hollow walling occur in the construction of the Parthenon. The wall was not an element of the building on which the Greek architect seemed to dwell with pleasure; much of it is almost invariably over-shadowed by the lines of columns which form the main features of the building.

The pediment (or gable) of a temple is a grand development of the walls, and perhaps the most striking of the additions which the Greeks made to the resources of the architect. It offers a fine field for sculpture, and adds real and apparent height beyond anything that the Egyptians ever attempted since the days of the Pyramid-builders; and it has remained in constant use to the present hour.

We do not hear of towers being attached to buildings, and, although such monumental structures as the Mausoleum of Halicarnassus approached the proportions of a tower, height does not seem to have commended itself to the mind of the Greek architect as necessary to the buildings which he designed. It was reserved for Roman and Christian art to introduce this element of architectural effect in all its power. On the other hand, the Greek, like the Persian architect, emphasised the base of his building in a remarkable manner, not only by base mouldings, but by planting the whole structure on a great range of steps which formed an essential part of the composition.

The Roof.

The construction of the roofs of Greek temples has been the subject of much debate. It is almost certain that they were in some way so made as to admit light. They were framed of timber and covered by tiles, often, if not always, of marble. Although all traces of the timber framing have disappeared, we can at least know that the pitch was not steep, by the slope of the outline of the pediments, which formed, as has already been said, perhaps the chief glory of a Greek temple. The flat stone roofs sometimes used by the Egyptians, and necessitating the placing of columns or other supports close together,

seem to have become disused, with the exception that
where a temple was surrounded by a range of columns
the space between the main wall and the columns was so
covered.

The vaulted stone roofs of the archaic buildings, of
which the treasury of Atreus (Figs. 52, 52a) was the
type, do not seem to have prevailed in a later period, or,
so far as we know, to have been succeeded by any similar
covering or vault of a more scientific construction.

It is hardly necessary to add that the Greek theatres
were not roofed. The Romans shaded the spectators in
their theatres and amphitheatres by means of a velarium
or awning, but it is extremely doubtful whether even
this expedient was in use in Greek theatres.

The Openings.

The most important characteristic of the openings in
Greek buildings is that they were flat-topped,—covered
by a lintel of stone or marble,—and never arched. We
have already * shown that this circumstance is really of
the first importance as determining the architectural
character of buildings. Doors and window openings were
often a little narrower at the top than the bottom, and
were marked by a band of mouldings, known as the architrave, on the face of the wall, and, so to speak, framing
in the opening. There was often also a small cornice over
each (Figs. 82, 83). Openings were seldom advanced
into prominence or employed as features in the exterior
of a building; in fact, the same effects which windows
produce in other styles were in Greek buildings created
by the interspaces between the columns.

* Chap. I.

The Columns.

These features, together with the superstructure or entablature, which they customarily carried, were the prominent parts of Greek architecture, occupying as they did the entire height of the building. The development of the orders (which we have explained to be really decorative systems, each of which involved the use of one sort of column, though the term is constantly understood as meaning merely the column and entablature) is a very interesting subject, and illustrates the acuteness with which the Greeks selected from those models which were accessible to them, exactly what was suited to their

FIG. 82.—GREEK DOORWAY, SHOWING CORNICE.

FIG. 83.—GREEK DOORWAY, FRONT VIEW. (FROM THE ERECHTHEIUM.)

purpose, and the skill with which they altered and refined, and almost redesigned, everything which they so selected.

During the whole period when Greek art was being developed, the ancient and polished civilisation of Egypt constituted a most powerful and most stable influence, always present,—always, comparatively speaking, within reach,—and always the same. Of all the forms of column and capital existing in Egypt, the Greeks, however, only selected that straight-sided fluted type of which the Beni-Hassan example is the best known, but by no means the only instance. We first meet with these fluted columns at Corinth, of very sturdy proportions, and having a wide, swelling, clumsy moulding under the abacus by way of a capital. By degrees the proportions of the shaft grew more slender, and the profile of the capital more elegant and less bold, till the perfected perfections of the Greek Doric column were attained. This column is the original to which all columns with moulded capitals that have been used in architecture, from the age of Pericles to our own, may be directly or indirectly referred; while the Egyptian types which the Greeks did not select—such, for example, as the lotus-columns at Karnak—have never been perpetuated.

A different temper or taste, and partly a different history, led to the selection of the West Asiatic types of column by a section of the Greek people; but great alterations in proportion, in the treatment of the capital, and in the management of the moulded base from which the columns sprang, were made, even in the orders which occur in the Ionic buildings of Asia Minor. This was carried further when the Ionic order was made use of in Athens herself, and as a result the Attic base and the perfected Ionic capital are to be found at their best in the Erechtheium example. The Ionic order and the Corinthian, which soon followed it, are the parents,—not, it is true, of all, but of the greater part of the columns with foliated capitals that have been used in all styles and periods of architecture since. It will not be forgotten that rude types of both orders are found represented on Assyrian bas-reliefs, but still the Corinthian capital and order must be considered as the natural and, so to speak, inevitable development of the Ionic. From the Corinthian capital an unbroken series of foliated capitals can be traced down to our own day; almost the only new ornamented type ever devised since being that which takes its origin in the Romanesque block capital, known to us in England as the early Norman cushion capital: this was certainly the parent of a distinct series, though even these owe not a little to Greek originals.

We have alluded to the Ionic base. It was derived from a very tall one in use at Persepolis, and we meet with it first in the rich but clumsy forms of the bases in the Asia Minor examples. In them we find the height of the feature as used in Persia compressed, while great, and to our eyes eccentric, elaboration marked the mouldings: these the refinement of Attic taste afterwards simplified, till the profile of the well-known Attic base was produced —a base which has had as wide and lasting an influence as either of the original forms of capital.

The Corinthian order, as has been above remarked, is the natural sequel of the Ionic. Had Greek architecture continued till it fell into decadence, this order would have been the badge of it. As it was, the decadence of Greek art was Roman art, and the Corinthian order was the favourite order of the Romans; in fact all the important examples of it which remain are Roman work.

If we remember how invariably use was made of one or other of the two great types of the Greek order in all the buildings of the best Greek time, with the addition towards its close of the Corinthian order, and that these orders, a little more subdivided and a good deal modified, have formed the substratum of Roman architecture and of that in use during the last three centuries; and if we also bear in mind that nearly all the columnar architecture of Early Christian, Byzantine, Saracenic, and Gothic times, owes its forms to the same great source, we may well admit that the invention and perfecting of the orders of Greek architecture has been—with one exception—the most important event in the architectural history of the world. That exception is, of course, the introduction of the Arch.

The Ornaments.

Greek ornaments have exerted the same wide influence over the whole course of Western art as Greek columns; and in their origin they are equally interesting as specimens of Greek skill in adapting existing types, and of Greek invention where no existing types would serve.

Few of the mouldings of Greek architecture are to be traced to anterior styles. There is nothing like them in Egyptian work, and little or nothing in Assyrian; and though a suggestion of some of them may no doubt be found in Persian examples, we must take them as having been substantially originated by Greek genius, which felt that they were wanted, designed them, and brought them far towards absolute perfection. They were of the most refined form, and when enriched were carved with consummate skill. They were executed, it must be remembered, in white marble,—a material having the finest surface, and capable of responding to the most delicate variations in contour by corresponding changes in shade or light in a manner and to a degree which no other material can equal. In the Doric, mouldings were few,

and almost always convex; they became much more numerous in the later styles, and then included many of
concave profile. The chief are the OVOLO, which formed
the curved part of the Doric capital, and the crowning
moulding of the Doric cornice; the CYMA; the BIRD'S BEAK,
employed in the capitals of the antæ; the FILLETS under
the Doric capital; the hollows and TORUS mouldings of
the Ionic and Corinthian bases.

The profiles of these mouldings were very rarely segments of circles, but lines of varying curvature, capable of
producing the most delicate changes of light and shade,
and contours of the most subtle grace. Many of them
correspond to conic sections, but it seems probable that
the outlines were drawn by hand, and not obtained by
any mechanical or mathematical method.

The mouldings were some of them enriched, to use the
technical word, by having such ornaments cut into them
or carved on them as, though simple in form, lent themselves well to repetition.* Where more room for ornament
existed, and especially in the capitals of the Ionic and
Corinthian orders, ornaments were freely and most grace-
fully carved, and very symmetrically arranged. Though
these were very various, yet most of them can be classed
under three heads. (1.) FRETS (Figs. 116 to 120). These
were patterns made up of squares or L-shaped lines inter-
laced and made to seem intricate, though originally simple.
Frequently these patterns are called Doric frets, from
their having been most used in buildings of the Doric
order. (2.) HONEYSUCKLE (Figs. 94 and 111 to 114). This
ornament, admirably conventionalised, had been used
freely by the Assyrians, and the Greeks only adopted

---

* For a statement of the general rule governing such enrichments,
see page 133.

what they found ready to their hand when they began
to use it; but they refined it, at the same time losing no
whit of its vigour or effectiveness, and the honeysuckle
has come to be known as a typical Greek decorative
motif. (3.) ACANTHUS (Figs. 84 and 85). This is a broad-
leaved plant, the foliage and stems of which, treated in a
conventional manner, though with but little departure
from nature, were found admirably adapted for floral
decorative work, and accordingly were made use of in the
foliage of the Corinthian capital, and in such ornaments
as, for example, the great finial which forms the summit
of the Choragic Monument of Lysicrates.

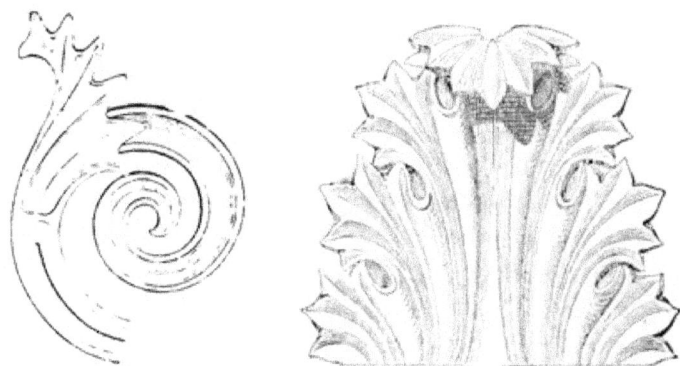

FIG. 84.—THE ACANTHUS LEAF AND STALK.

The beauty of the carving was, however, eclipsed by
that highest of all ornaments—sculpture. In the Doric
temples, as, for example, in the Parthenon, the architect
contented himself with providing suitable spaces for the
sculptor to occupy; and thus the great pediments, the
metopes (Fig. 86) or square panels, and the frieze of the
Parthenon were occupied by sculpture, in which there
was no necessity for more conventionalism than the
amount of artificial arrangement needed in order fitly

to occupy spaces that were respectively triangular, square, or continuous. In the later and more voluptuous style of the Ionic temples we find sculpture made into an architectural feature, as in the famous statues, known as the Caryatides, which support the smallest portico of the Erechtheium, and in the enriched columns of the Temple of Diana at Ephesus. Sculpture had already been so employed in Egypt, and was often so used in later times; but the best opportunity for the display of the finest qualities of the sculptor's art is such an one as the pediments, &c., of the great Doric temples afforded.

There is little room for doubting that all the Greek temples were richly decorated in colours, but traces and indications are all that remain: these, however, are sufficient to prove that a very large amount of colour was employed, and that probably ornaments (Figs. 105 to 120) were painted upon many of those surfaces which were left plain by the mason, especially on the cornices, and that mosaics (Fig. 87) and coloured marbles, and even gilding, were freely used. There is also ground for believing that as the use of carved enrichments increased with the increasing adoption of the Ionic and Corinthian styles, less use was made of painted decorations.

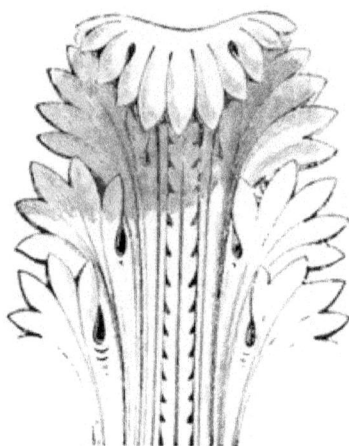

FIG. 5.—THE ACANTHUS LEAF.

K

Architectural Character.

Observations which have been made during the course of
this and the previous chapters will have gone far to point
out the characteristics of Greek art. An archaic and
almost forbidding severity, with heavy proportions and

FIG. 86.—METOPE FROM THE PARTHENON. CONFLICT BETWEEN
A CENTAUR AND ONE OF THE LAPITHÆ.

more strength than grace, marks the earliest Greek buildings of which we have any fragments remaining. Dignity,
sobriety, refinement, and beauty are the qualities of the
works of the best period. The latest buildings were
more rich, more ornate, and more slender in their proportions, and to a certain extent less severe.

Most carefully studied proportions prevailed, and were
wrought out to a pitch of completeness and refinement
which is truly astounding. Symmetry was the all but

FIG. 87.—MOSAIC FROM THE TEMPLE OF ZEUS, OLYMPIA.

invariable law of composition. Yet in certain respects—
as, for example, the spacing and position of the columns—
a degree of freedom was enjoyed which Roman archi-

K 2

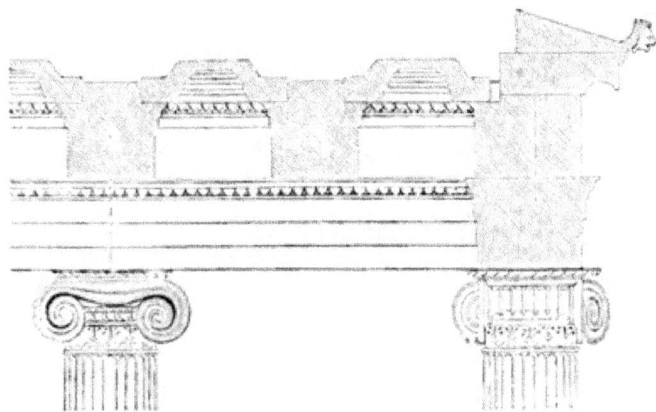

FIG. 88.—SECTION OF THE PORTICO OF THE ERECHTHEUM.

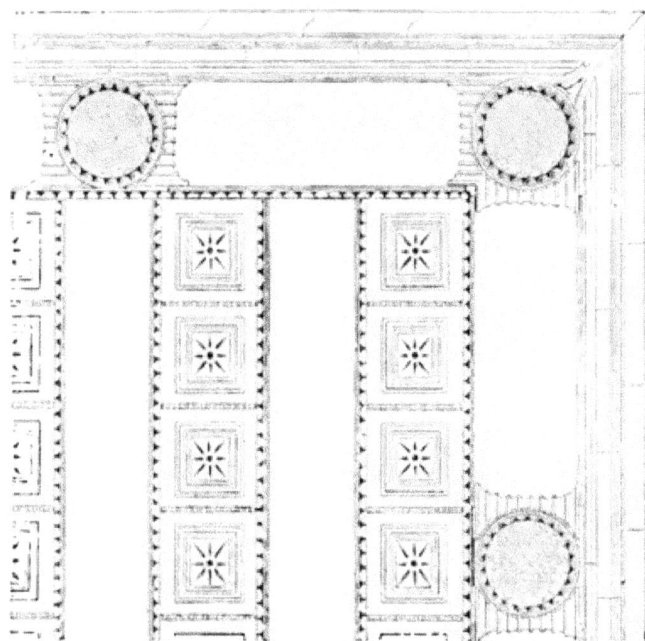

FIG. 89.—PLAN OF THE PORTICO—LOOKING UP.

EXAMPLES OF GREEK ORNAMENT
IN THE NORTHERN PORTICO OF THE ERECHTHEUM—SHOWING THE ORNAMENTATION
OF THE CEILING.

tecture did not possess. Repetition ruled to the almost
entire suppression of variety. Disclosure of the arrangement and construction of the building was almost complete, and hardly a trace of concealment can be detected.
Simplicity reigns in the earliest examples; the elaboration of even the most ornamental is very chaste and
graceful; and the whole effect of Greek architecture is
one of harmony, unity, and refined power.

A general principle seldom pointed out which governs
the application of enrichments to mouldings in Greek
architecture may be cited as a good instance of the subtle
yet admirable concord which existed between the different
features: it is as follows. *The outline of each enrichment in
relief was ordinarily described by the same line as the profile
of the moulding to which it was applied.* The egg enrichment (Fig. 91) on the ovolo, the water-leaf on the cyma
reversa (Figs. 92 and 97), the honeysuckle on the cyma
recta (Fig. 94), and the guilloche (Fig. 100) on the torus,
are examples of the application of this rule,—one which
obviously tends to produce harmony.

FIG. 90.—CAPITAL OF ANTÆ FROM THE ERECHTHEIUM.

FIG. 91.—EGG AND DART.

FIG. 92.—LEAF AND DART.

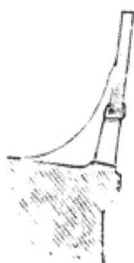

FIG. 93.—HONEYSUCKLE.

FIG. 94.—HONEYSUCKLE.

FIG. 95.—ACANTHUS.

FIG. 96.—ACANTHUS.

EXAMPLES OF GREEK ORNAMENT IN RELIEF.

FIG. 97.—LEAF AND TONGUE.

FIG. 98.—LEAF AND TONGUE.

FIG. 99.—GARLAND

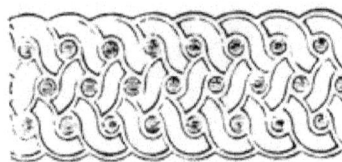

FIG. 101.—BEAD AND FILLET.

FIG. 102.—BEAD AND FILLET.

FIG. 100.—GUILLOCHE.

FIG. 103.—TORUS MOULDING.

FIG. 104.—TORUS MOULDING.

EXAMPLES OF GREEK ORNAMENT IN RELIEF.

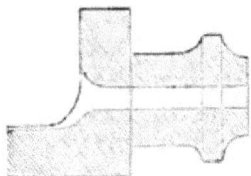

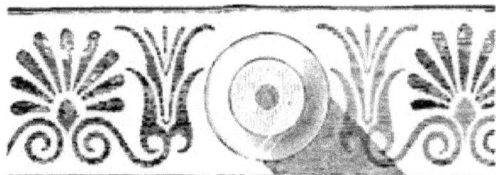

FIG. 105.—HONEYSUCKLE.

FIG. 107.—HONEYSUCKLE.

FIG. 106.

FIG. 108.

FIGS. 106, 108.—FACIAS WITH BANDS OF FOLIAGE.

FIG. 109.—LEAF AND DART.

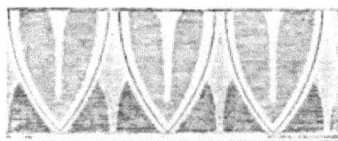

FIG. 110.—EGG AND DART.

EXAMPLES OF GREEK ORNAMENT IN COLOUR.

FIG. 111.

FIG. 112.

FIG. 113.

FIG. 114.

FIG. 115.—GUIL-
LOCHE.

FIGS. 111 TO 113.—EXAMPLES OF THE HONEYSUCKLE.

FIG. 114.—COMBINATION OF THE FRET, THE EGG AND DART, THE BEAD
AND FILLET, AND THE HONEYSUCKLE.

FIG. 118.

FIG. 119.

FIG. 120.

FIGS. 116 TO 120.—EXAMPLES OF THE FRET.

EXAMPLES OF GREEK ORNAMENT IN COLOUR.

Fig. 121.—Elevation of an Etruscan Temple (restored from descriptions only).

CHAPTER VIII.

ETRUSCAN AND ROMAN ARCHITECTURE.

Historical and General Sketch.

THE few grains of truth that we are able to sift from the mass of legend which has accumulated round the early history of Rome seem to indicate that at a very early period—which the generally received date of 753 B.C. may be taken to fix as nearly as is now possible—a small band of outcasts and marauders settled themselves on the Palatine Hill and commenced to carry on depredations against the various cities of the tribes whose territories were in the immediate neighbourhood, such as the Umbrians, Sabines, Samnites, Latins, and Etruscans. A walled city was built, which from its admirable situation succeeded in attracting inhabitants in considerable

numbers, and speedily began to exercise supremacy over
its neighbours. The most important of the neighbouring nations were the Etruscans, who called themselves
Rasena, and who must have settled on the west coast of
Italy, between the rivers Arno and Tiber, at a very early
period. Their origin is, however, very obscure, some
authorities believing, upon apparently good grounds, that
they came from Asia Minor, while others assert that they
descended from the north over the Rhætian Alps. But
whatever that origin may have been, they had at the time
of the founding of Rome as a city attained a high degree
of civilisation, and showed a considerable amount of
architectural skill; and their arts exercised a very great
influence upon Roman art.

Considerable remains of the city walls of several Etruscan towns still exist. These show that the masonry was
of what has been termed a Cyclopean character,—that is
to say, the separate stones were of an enormous size; in
the majority of examples these stones were of a polygonal
shape, though in a few instances they were rectangular,
while in all cases they were fitted together with the most
consummate accuracy of workmanship, which, together
with their great massiveness, has enabled much of this
masonry to endure to the present day. Cortona, Volterra,
Fiesole, and other towns exhibit instances of this walling.
The temples, palaces, or dwelling-houses which went to
make up the cities so fortified have all disappeared, and
the only existing structural remains of Etruscan buildings are tombs. These are found in large numbers, and
consist—as in the earlier instances which have already
been described—both of rock-cut and detached erections.
Of the former, the best known group is at Castel d'Asso,
where we find not only chambers cut into the rock, each

resembling an ordinary room with an entrance in the
face of the rock, but also monuments cut completely
out and standing clear all round; and we cannot fail to
detect in the forms into which the rock has been cut,

FIG. 122.—SEPULCHRE AT CORNETO.

especially those of the roof, imitations of wooden build
ings, heavy square piers being left at intervals supporting longitudinal beams which hold up the roof.
Fig. 122 is an illustration of the interior of a chamber
in the rock. Occasionally there were a cornice and pediment over the entrance.

The other class of tombs are circular tumuli, similar
to the Pelasgic tombs of Asia Minor; of these large
numbers exist, but not sufficiently uninjured to enable
us to restore them completely. They generally consisted
of a substructure of stone, upon which was raised a
conical elevation. In the case of the Regulini Galeassi
tomb there were an inner and an outer tumulus, the
latter of which covered several small tombs, while the
inner enclosed one only, which had fortunately never
been opened till it was lately discovered. This tomb
was vaulted on the horizontal system—that is to say,
its vault was not a true arch, but was formed of courses
of masonry, each overhanging the one below, as in the
Treasury of Atreus, and it had a curious recess in the
roof, in which were found numerous interesting examples
of Etruscan pottery. It is, however, clear from the city
gates, sewers, aqueducts, &c., that the Etruscans were
acquainted with and extensively used the true radiating
arch composed of wedge-shaped stones (voussoirs), and
that they constructed it with great care and scientific
skill. The gate at Perugia, and the Cloacæ or Sewers
at Rome, constructed during the reign of the Tarquins,*
at the beginning of the sixth century B.C., are examples of
the true arch, and this makes it certain that it was from
the Etruscans that the Romans learned the arched construction which, when combined with the trabeated or
lintel mode of construction which they copied from the
Greeks, formed the chief characteristic of Roman architecture. The Cloaca Maxima (Fig. 123), which is roofed
over with three concentric semicircular rings of large

* The story of the Tarquins probably points to a period when the
chief supremacy at Rome was in the hands of an Estruscan family, and is
interesting for this reason.

stones, still exists in many places with not a stone displaced, as a proof of the skill of these early builders. There are remains of an aqueduct at Tusculum which are interesting from the fact of the horizontal being combined with the true arch in its construction.

No Etruscan temples remain now, but we know from Vitruvius that they consisted of three cells with one or more rows of columns in front, the intercolumniation or interval between the columns being excessive. The

Fig. 123.—CLOACA MAXIMA.

largest Etruscan temple of which any record remains was that of Jupiter Capitolinus at Rome, which, under the Empire, became one of the most splendid temples of antiquity. It was commenced by Tarquinius Superbus, and is said to have derived its name from the fact of the builders, when excavating the foundations, coming upon a freshly bleeding head (*caput*), indicating that the place would eventually become the chief city of the world. Another form of Etruscan temple is described by Vitruvius, consisting of one circular cell only, with a porch. This form was probably the origin of the series of circular Roman buildings which includes such forms of temples as that at Tivoli, and many of the famous mansolea, *e.g.* that of Hadrian, and the culmination of which style is seen in the Pantheon. It is interesting to notice that the Romans never entirely gave up the circular form, one instance of its use in Britain at a late period of the Roman occupation having been discovered in the ruins of Silchester near Basingstoke; and we shall find that it was perpetuated in Christian baptisteries, tombs, and occasionally churches.

We know from the traces of such buildings which exist, that the Etruscans must have constructed theatres and amphitheatres, and it is recorded that the first Tarquin laid out the Circus Maximus and instituted the great games held there. At Sutri there are ruins of an amphitheatre which is nearly a perfect circle, measuring 265 ft. in its greatest breadth and 295 ft. in length.

There are no remains of other buildings which would enable us to form an opinion as to the civic architecture of the Etruscans: they must, however, have attained to a considerable skill in sculpture, as in some of the tombs figures are represented in high relief which show no small power of expression. They, too, like the Egyptians, embellished their tombs with mural paintings. These are generally in outline, and represent human figures and animals in scenes of every-day life, with conventionalised foliage, or mythological scenes such as the passage of the soul after death to the judgment-seat where its actions in life are to be adjudicated upon. In the plastic arts the Etruscans made great progress, many of their vases showing a delicacy and grace which have

never been surpassed, and exhibiting in their decorations
traces of both Greek and Egyptian influence.

We now reach the last of the classical styles of antiquity, the Roman,—a style which, however, is rather an
adaptation or amalgamation of other styles than an original
and independent creation or development. The contrast
is very great between the "lively Grecian," imaginative
and idealistic in the highest degree—who seemed to have
an innate genius for art and beauty, and who was always
eager to perpetuate in marble his ideal conception of the
"hero from whose loins he sprung," or to immortalise
with some splendid work of art the name of his mother-
city—and the stern, practical Roman, realistic in his every
pore, eager for conquest, and whose one dominant idea was
to bring under his sway all the nations who were brought
into contact with him, and to make his city—as had been
foretold—the capital of the whole world. With this idea
always before him, it is no wonder that such a typical
Roman as M. Porcius Cato should look with disdain upon
the fine arts in all their forms, and should regard a love
for the beautiful, whether in literature or art, as synonymous with effeminacy. Mummius, also, who destroyed
Corinth, is said to have been so little aware of the value
of the artistic treasures which he carried away, as to
stipulate with the carriers who undertook to transport
them to Rome, that if any of the works of art were lost
they should be replaced by others of equal value.

When the most prominent statesmen displayed such
indifference, it is not surprising that for nearly 500 years
no single trace of any architectural building of any merit
at all in Rome can now be discovered, and that history is
silent as to the existence of any monuments worthy of

being mentioned. Works of public utility of a very extensive nature were indeed carried out during this period;
such, for example, as the Appian Way from Rome to
Capua, which was the first paved road in Rome, and was
constructed by the Censor Appius Claudius in B.C. 309.
This was 14 ft. wide and 3 ft. thick, in three layers:
1st, of rough stones grouted together; 2nd, of gravel; and
3rd, of squared stones of various dimensions. The same
Censor also brought water from Præneste to Rome by a
subterranean channel 11 miles long. Several bridges were
also erected, and Cato the Censor is said to have built a
basilica.

Until about 150 B.C. all the buildings of Rome were
constructed either of brick or the local stone; and though
we hear nothing of architecture as a fine art, we
cannot hesitate to admit that during this period the
Romans carried the art of construction, and especially
that of employing materials of small dimensions and
readily obtainable, in buildings of great size, to a remarkable pitch of perfection. It was not till after the fall of
Carthage and the destruction of Corinth, when Greece
became a Roman province under the name of Achaia—
both which events occurred in the year 146 B.C.—that
Rome became desirous of emulating, to a certain extent,
the older civilisation which she had destroyed; and about
this time she became so enormously wealthy that vast
sums of money were expended, both publicly and privately,
in the erection of monuments, many of which remain to
the present day, more or less altered. The first marble
temple in Rome was built by the Consul Q. Metellus
Macedonicus, who died B.C. 115. Roman architecture from
this period began to show a wonderful diversity in the
objects to which it was directed,—a circumstance perhaps

L

as interesting as its great scientific and structural advance upon all preceding styles. In the earlier styles temples, tombs, and palaces were the only buildings deemed worthy of architectural treatment; but under the Romans baths, theatres, amphitheatres, basilicas, aqueducts, triumphal arches, &c., were carried out just as elaborately as the temples of the gods.

It was under the Emperors that the full magnificence of Roman architectural display was reached. The famous boast of Augustus, that he found Rome of brick and left her of marble, gives expression in a few words to what was the great feature of his reign. Succeeding emperors lavished vast sums on buildings and public works of all kinds; and thus it comes to pass that though the most destructive of all agencies, hostile invasions, conflagrations, and long periods of neglect, have each in turn done their utmost to destroy the vestiges of Imperial Rome, there still remain fragments, and in one or two instances whole monuments, enough to make Rome, after Athens, the richest store of classic architectural antiquities in the world.

But it was not in Rome only that great buildings were erected. The whole known civilised world was under Roman dominion, and wherever a centre of government or even a flourishing town existed there sprang up the residence of the dominant race, and their places of business, public worship, and public amusement. Consequently, we find in our own country, and in France, Spain, Germany, Italy, North Africa, and Egypt—in short, in all the countries where Roman rule was established—examples of temples, amphitheatres, theatres, triumphal arches, and dwelling-houses, some of them of great interest and occasionally in admirable preservation.

Fig. 124.—"Incantada" in Salonica.

CHAPTER IX.

THE BUILDINGS OF THE ROMANS.

THE temples in Rome were not, as in Greece and
Egypt, the structures upon which the architect
lavished all the resources of his art and his science. The
general form of them was copied from that made use
of by the Greeks, but the spirit in which the original
idea was carried out was entirely different. In a word,
the temples of Rome were by no means worthy of her
size and position as the metropolis of the world, and
very few remains of them exist.

Ten columns are still standing of the Temple of Antoninus and Faustina (now the church of San Lorenzo in
Miranda): it occupied the site of a previous temple
and was dedicated by Antoninus Pius to his wife Faustina. The Temple (supposed) of Fortuna Virilis, in the
Ionic style (Fig. 125), still exists as the church of Santa
Maria Egiziaca: this was tetrastyle, with half-columns
all round it, and this was of the kind called by Vitruvius "pseudo-peripteral." A few fragmentary remains of

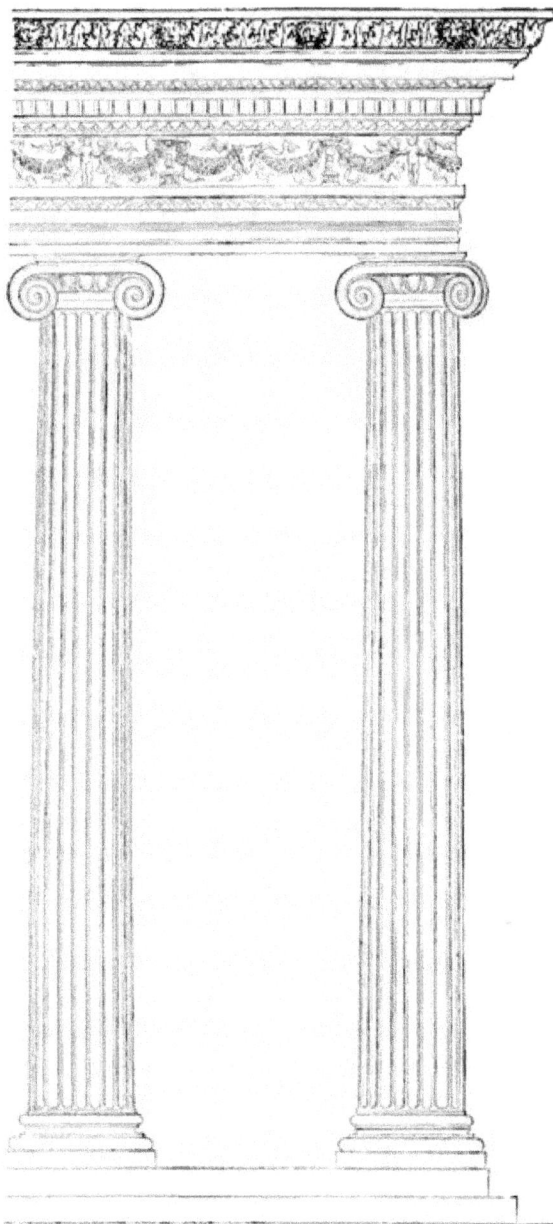

FIG. 125.—IONIC ORDER FROM THE TEMPLE OF FORTUNA VIRILIS, ROME.

other temples exist in Rome, but in some of the Roman
provinces far finer specimens of temples remain, of
which perhaps the best is the Maison Carrée at Nimes
(Fig. 126). Here we find the Roman plan of a single
cell and a deep portico in front, while the sides and
rear have the columns attached. The intercolumnia-
tions and the details of the capitals and entablature are,
however, almost pure Greek. The date of this temple is
uncertain, but it is most probable that it was erected
during the reign of Hadrian. The same emperor is
said to have completed the magnificent Temple of Jupiter
Olympius at Athens, which was 354 ft. long by 171 ft. wide.
It consisted of a cell flanked on each side by a double row
of detached columns; in front was one row of columns in
antis, and three other rows in front of these, while there
were also three rows in the rear: as the columns were of
the Corinthian order, and nearly 60 ft. in height, it may
be imagined that it was a splendid edifice.

The ruins of another magnificent provincial Roman
temple exist at Baalbek—the ancient Heliopolis—in Syria,
not far from Damascus. This building was erected during
the time of the Antonines, probably by Antoninus Pius
himself, and originally it must have been of very extensive dimensions, the portico alone being 180 ft. long
and about 37 ft. deep. This gives access to a small
hexagonal court, on the western side of which a triple
gateway opens into the Great Court, which is a vast
quadrangle about 450 ft. long by 400 ft. broad, with
ranges of small chambers or niches on three sides, some
of which evidently had at one time beautifully groined
roofs. At the western end of this court, on an artificial
elevation, stand the remains of what is called the Great
Temple. This was originally 290 ft. long by 160 ft.

FIG. 126.—ROMAN-CORINTHIAN TEMPLE AT NIMES (MAISON-CARRÉE). PROBABLY OF THE TIME OF HADRIAN.

wide, and had 54 columns supporting its roof, six only
of which now remain erect. The height of these columns,
including base and capital, is 75 ft., and their diameter
is 7 ft. at base and about 6 ft. 6 in. at top; they are
of the Corinthian order, and above them rises an elaborately moulded entablature, 14 ft. in height. Each of
the columns is composed of three stones only, secured by
strong iron cramps; and indeed one of the most striking
features of this group of buildings is the colossal size
of the stones used in their construction. The quarries
from which these stones were hewn are close at hand,
and in them is one stone surpassing all the others in
magnitude, its dimensions being 68 ft. by 14 ft. 2 in. by
13 ft. 11 in. It is difficult to imagine what means can
have existed for transporting so huge a mass, the weight
of which has been calculated at 1100 tons.

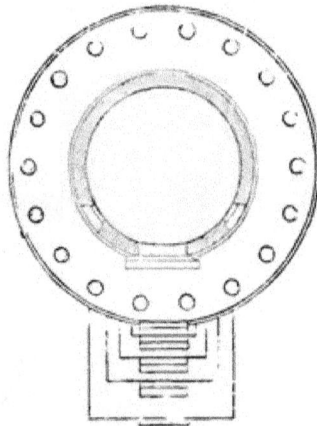

FIG. 127.—GROUND-PLAN OF THE TEMPLE OF VESTA AT TIVOLI.

Other smaller temples exist in the vicinity, all of
which are lavishly decorated, but on the whole the

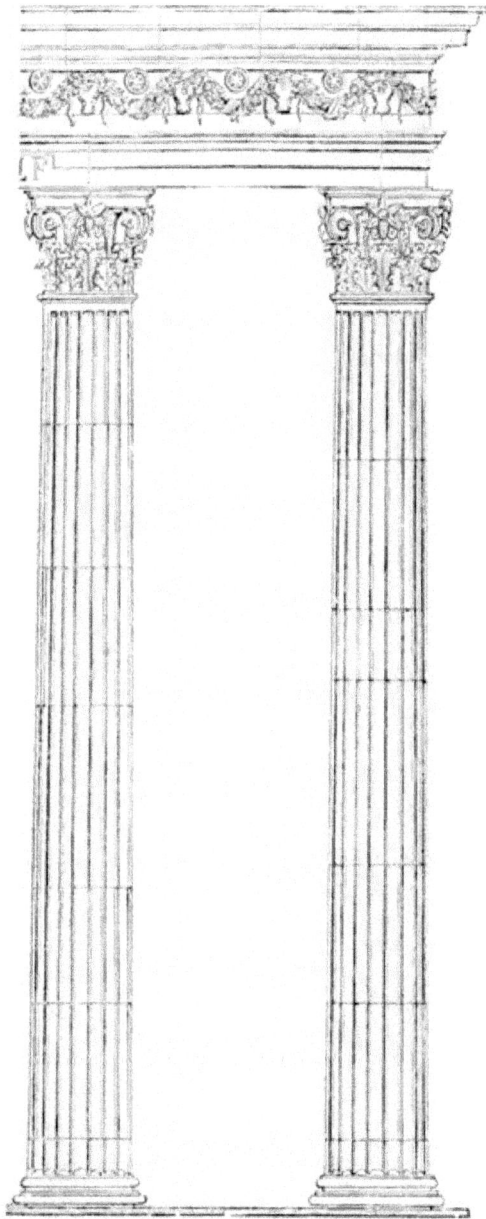

FIG. 128.—CORINTHIAN ORDER FROM THE TEMPLE OF VESTA AT TIVOLI.

ornamentation shows an exuberance of detail which
somewhat offends a critical artistic taste.

Circular temples were an elegant variety, which seems

FIG. 129.—THE TEMPLE OF VESTA AT TIVOLI. PLAN (LOOKING UP) AND SECTION
OF PART OF THE PERISTYLE.

to have been originated by the Romans, and of which
two well-known examples remain—the Temples of Vesta
at Rome and at Tivoli. The columns of the temple

at Tivoli (Fig. 128) form a well-known and pleasing
variety of the Corinthian order, and the circular form
of the building as shown on the plan (Fig. 127) gives
excellent opportunities for good decorative treatment,
as may be judged of by the enlarged diagram of part
of the peristyle (Fig. 129).

Basilicas.

Among the most remarkable of the public buildings of
Roman times, both in the mother-city and in the provinces,
were the Basilicas or Halls of Justice, which were also
used as commercial exchanges. It is also believed that
Basilicas existed in some Greek cities, but no clue to their
structural arrangements exists, and whence originated the
idea of the plan of these buildings we are unable to state;
their striking similarity to some of the rock-cut halls or
temples of India has been already pointed out. They
were generally (though not always) covered halls, oblong
in shape, divided into three or five aisles by two or more
rows of columns, the centre aisle being much wider than
those at the sides: over the latter, galleries were frequently
erected. At one end was a semicircular recess or apse, the
floor of which was raised considerably above the level of
the rest of the building, and here the presiding magistrate
sat to hear causes tried. Four * of these buildings are
mentioned by ancient writers as having existed in republican times, viz. the Basilica Portia, erected in B.C. 184, by
Cato the Censor; the Basilica Emilia et Fulvia, erected in
B.C. 179 by the censors M. Fulvius Nobilior and M. Æmilius
Lepidus, and afterwards enlarged and called the Basilica

* The passage in Varro, which is the sole authority for the Basilica
Opimia, is generally considered to be corrupt.

Paulli; the Basilica Sempronia, erected in B.C. 169 by
Tib. Sempronius Gracchus; and the Basilica Julia, erected
by Julius Cæsar, B.C. 46. All these buildings had wooden
roofs, and were of no great architectural merit, and they
perished at a remote date. Under the Empire, basilicas
of much greater size and magnificence were erected; and
remains of that of Trajan, otherwise called the Basilica
Ulpia, have been excavated in the Forum of Trajan.
This was about 360 ft. long by 180 ft. wide, had four rows

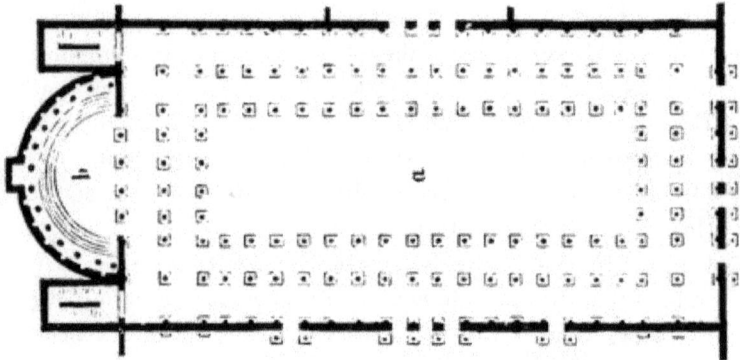

FIG. 130.—GROUND-PLAN OF THE BASILICA ULPIA, ROME.

of columns inside, and is supposed to have been covered
by a semicircular wooden roof. Apollodorus of Damascus
was the architect of this building. Another basilica of
which remains exist is that of Maxentius, which, after his
overthrow by Constantine in A.D. 312, was known as the
Basilica Constantiniana. This structure was of stone, and
had a vaulted roof; it was 195 ft. between the walls, and
was divided into three aisles by piers with enormous
columns standing in front of them.

One provincial basilica, that at Trèves, still stands;
and although it must have been considerably altered, it

is by far the best existing example of this kind of building. The internal columns do not exist here, and it is simply a rectangular hall about 175 ft. by 85 ft., with the usual semicircular apse.

The chief interest attaching to these basilicas lies in the fact that they formed the first places of Christian assembly, and that they served as the model upon which the first Christian churches were built.

Theatres and Amphitheatres.

Although dramas and other plays were performed in Rome as early as 240 B.C., there seems to have been a strong prejudice against permanent buildings for their representation, as it is recorded that a decree was passed in B.C. 154 forbidding the construction of such buildings. Mummius, the conqueror of Corinth, obtained permission to erect a wooden theatre for the performance of dramas as one of the shows of his triumph, and after this many buildings of the kind were erected, but all of a temporary nature; and it was not till B.C. 61 that the first permanent theatre was built by Pompey. This, and the theatres of Balbus and Marcellus, appear to have been the only permanent theatres that were erected in Imperial Rome; and there are no remains of any but the last of these, and this is much altered. So that, were it not for the remains of theatres found at Pompeii, it would be almost impossible to tell how they were arranged; but from these we can see that the stage was raised and separated from the part appropriated to the spectators by a semicircular area, much like that which in Greek theatres was allotted to the chorus: in the Roman ones this was assigned for the use of the senators. The portion devoted to the spectators—called the Cavea—was also semicircular on plan, and consisted of tiers of steps rising one above the other, and divided at intervals by wide passages and converging staircases communicating with the porticoes, which ran round the whole theatre at every story.

At Orange, in the South of France, are the remains of a very fine theatre, similar in plan to that described. The

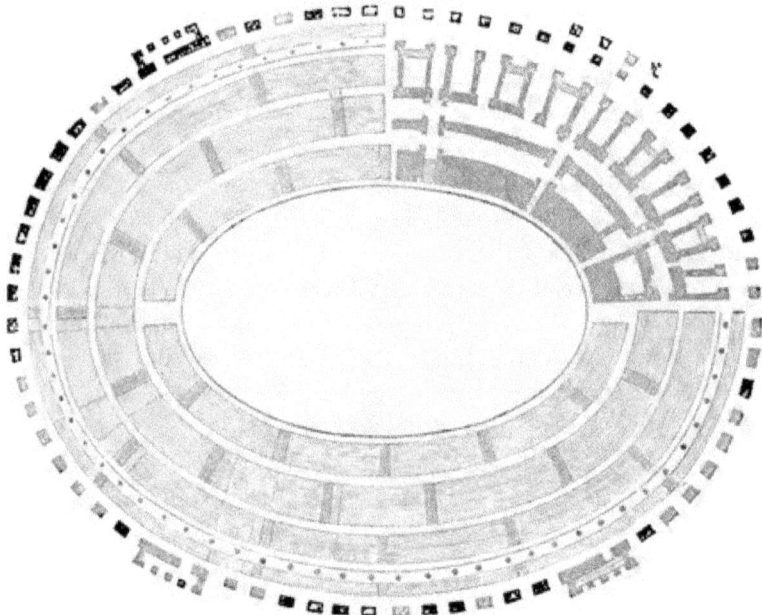

Fig. 131.—Plan of the Colosseum, Rome.

great wall which formed the back of the scene in this building is still standing, and is one of the most magnificent pieces of masonry existing.

Although the Romans were not particularly addicted

to dramatic representations, yet they were passionately
fond of shows and games of all kinds: hence, not only in
Rome itself, but in almost every Roman settlement, from
Silchester to Verona, are found traces of their amphitheatres, and the mother-city can claim the possession of

FIG. 132.—THE COLOSSEUM. SECTION AND ELEVATION.

the most stupendous fabric of the kind that was ever
erected—the Colosseum or Flavian Amphitheatre, which
was commenced by Vespasian and finished by his son
Titus. An amphitheatre is really a double theatre with-

out a stage, and with the space in the centre unoccupied by seats. This space, which was sunk several feet below the first row of seats, was called the arena, and was appropriated to the various exhibitions which took place in the building. The plan was elliptical or oval, and this shape seems to have been universal.

The Colosseum, whose ruins still remain to attest its pristine magnificence—

> " Arches on arches, as it were that Rome,
> Collecting the chief trophies of her line,
> Would build up all her triumphs in one dome "*—

was 620 ft. long and 513 wide, and the height was about 162 ft. It was situated in the hollow between the Esquiline and Cælian hills. The ranges of seats were admirably planned so as to enable all the audience to have a view of what was going on in the arena, and great skill was shown both in the arrangement of the approaches to the different tiers and in the structural means for supporting the seats, and double corridors ran completely round the building on each floor, affording ready means of exit. Various estimates have been made of the number of spectators that could be accommodated, and these range from 50,000 to 100,000, but probably 80,000 was the maximum. Recent excavations have brought to light the communications which existed between the arena and the dens where the wild animals and human slaves and prisoners were confined, and some of the water channels used when mimic sea-fights were exhibited. The external façade is composed of four stories, separated by entablatures that run completely round the building without a break. The three lower stories consist of a series of semicircular arched openings, eighty

* Byron.

in number, separated by piers with attached columns in front of them, the Doric order being used in the lowest story, the Ionic in the second, and the Corinthian in the third; the piers and columns are elevated on stylobates; the entablatures have a comparatively slight projection, and there are no projecting keystones in the arches. In the lowest range these openings are 13 ft. 4 in. wide, except the four which are at the ends of the two axes of the ellipse, and these are 14 ft. 6 in. wide. The diameter of the columns is 2 ft. 8¾ in. The topmost story, which is considerably more lofty than either of the lower ones, was a nearly solid wall enriched by Corinthian pilasters. In this story occur two tiers of small square openings in the alternate spaces between the pilasters. These openings are placed accurately over the centres of the arches of the lower stories. Immediately above the higher range of square openings are a series of corbels—three between each pair of pilasters—which probably received the ends of the poles carrying the huge awning which protected the spectators from the sun's rays. The whole is surmounted by a heavy cornice, in which, at intervals immediately over each corbel, are worked square mortise holes, forming sockets through which the poles of the awning passed. The stone of which the façade of the Colosseum is built is a local stone, called travertine, the blocks of which are secured by iron cramps without cement. Nearly all the internal portion of the building is of brick, and the floors of the corridors, &c., are paved with flat bricks covered with hard stucco. These amphitheatres were occasionally the scene of imitations of marine conflicts, when the arena was flooded with water and mimic vessels of war engaged each other. Very complete arrangements were made, by means of small aqueducts, for leading the water into the arena and for carrying it off.

Apart from theatrical representations and gladiatorial combats, the Romans had an inordinate passion for chariot races. For these the circi were constructed, of which class of buildings the Circus Maximus was the largest. This, originally laid out by Tarquinius Priscus, was reconstructed on a larger scale by Julius Cæsar. It was circular at one end and rectangular at the other, at which was the entrance. On both sides of the entrance were a number of small arched chambers, called *carceres*, from which the chariots started. The course was divided down the centre by a low wall, called the *spina*, which was adorned with various sculptures. The seats rose in a series of covered porticoes all round the course, except at the entrance. As the length of the Circus Maximus was nearly 700 yards, and the breadth about 135 yards, it is possible that Dionysius may not have formed an exaggerated notion of its capacity when he says it would accommodate 150,000 spectators.

In the Roman provinces amphitheatres were often erected; and at Pola in Istria, Verona in Italy, and Nimes and Arles in France, fine examples remain. A rude Roman amphitheatre, with seats cut in the turf of a hill-side, exists to this day at the old town of Dorchester in Dorset, which was anciently a Roman settlement.

Baths (Thermæ).

Nothing can give us a more impressive idea of the grandeur and lavish display of Imperial Rome than the remains of the huge Thermæ, or bathing establishments, which still exist. Between the years 10 A.D., when

M

Agrippa built the first public baths, and 324 A.D., when those of Constantine were erected, no less than twelve of these vast establishments were erected by various emperors, and bequeathed to the people. Of the whole number, the baths of Caracalla and of Diocletian are the only ones which remain in any state of preservation, and these were probably the most extensive and magnificent of all. All these splendid buildings were really nothing more than bribes to secure the favour of the populace; for it seems quite clear that the public had practically free entrance to them, the only charge mentioned by writers of the time being a quadrans, about a farthing of our money. Gibbon says, "The meanest Roman could purchase with a small copper coin the daily enjoyment of a scene of pomp and luxury which might excite the envy of the kings of Asia." And this language is not exaggerated. Not only were there private bath-rooms, swimming-baths, hot baths, vapour baths, and, in fact, all the appurtenances of the most approved Turkish baths of modern times, but there were also gymnasia, halls for various games, lecture-halls, libraries, and theatres in connection with the baths, all lavishly ornamented with the finest paintings and sculpture that could be obtained. Stone seems to have been but sparingly used in the construction of these buildings, which were almost entirely of brick faced with stucco: this served as the ground for an elaborate series of fresco paintings.

The baths of Caracalla, at the foot of the Aventine hill, erected A.D. 217, comprised a quadrangular block of buildings of about 1150 ft. (about the fifth of a mile) each way. The side facing the street consisted of a portico the whole length of the façade, behind which were numerous ranges of private bath-rooms. The side and rear blocks contained

numerous halls and porticoes, the precise object of which
it is now very difficult to ascertain. As Byron says :

> " Temples, baths, or halls?
> Pronounce who can."

This belt of buildings surrounded an open courtyard or
garden, in which was placed the principal bathing establishment (Fig. 133), a building 730 ft. by 380 ft., which

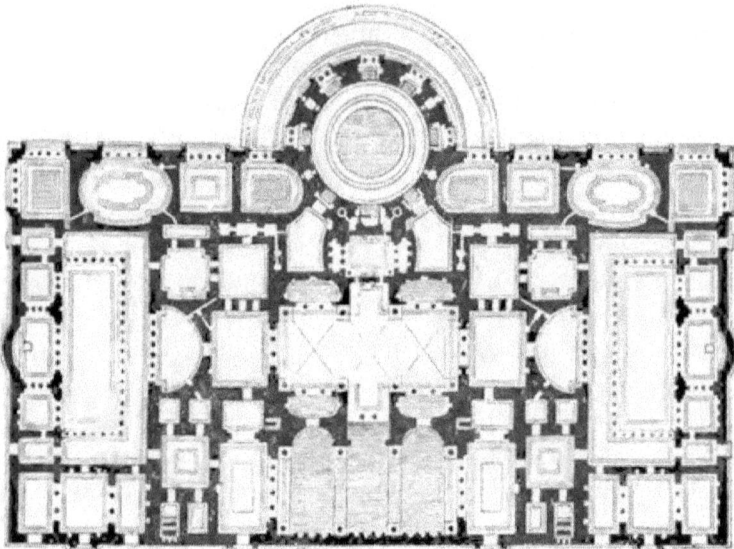

Fig. 133.—Plan of the Principal Building, Baths of Caracalla, Rome.

contained the large piscina, or swimming-bath, various hot
baths, dressing-rooms, gymnasia, and other halls for athletic
exercises. In the centre of one of the longer sides was
a large semicircular projection, roofed with a dome, which
was lined with brass : this rotunda was called the solar

cell. From the ruins of these baths were taken some of the most splendid specimens of antique sculpture, such as the Farnese Hercules and the Flora in the Museum of Naples.

The baths of Diocletian, erected just at the commencement of the fourth century A.D., were hardly inferior to those of Caracalla, but modern and ancient buildings are now intermingled to such an extent that the general plan of the buildings cannot now be traced with accuracy. There are said to have been over 3000 marble seats in these baths; the walls were covered with mosaics, and the columns were of Egyptian granite and green Numidian marble. The Ephebeum, or grand hall, still exists as the church of Santa Maria degli Angeli, having been restored by Michelangelo. It is nearly 300 ft. long by 90 ft. wide, and is roofed by three magnificent cross vaults, supported on eight granite columns 45 ft. in height. (Fig. 134.)

There is one ancient building in Rome more impressive than any other, not only because it is in a better state of preservation, but because of the dignity with which it has been designed, the perfection with which it has been constructed, and the effectiveness of the mode in which its interior is lighted. We allude to the Pantheon. Opinions differ as to whether this was a Hall attached to the thermæ of Agrippa, or whether it was a temple. Without attempting to determine this point, we may at any rate claim that the interior of this building admirably illustrates the boldness and telling power with which the large halls forming part of the thermæ were designed; and, whether it belonged to such a building or not, it is wonderfully well fitted to illustrate this subject.

The Pantheon is the finest example of a domed hall

which we have left. The building, which forms the church of Santa Maria ad Martyres, has been considerably altered at various times since its erection, and now consists of a rotunda with a rectangular portico in front of it. The

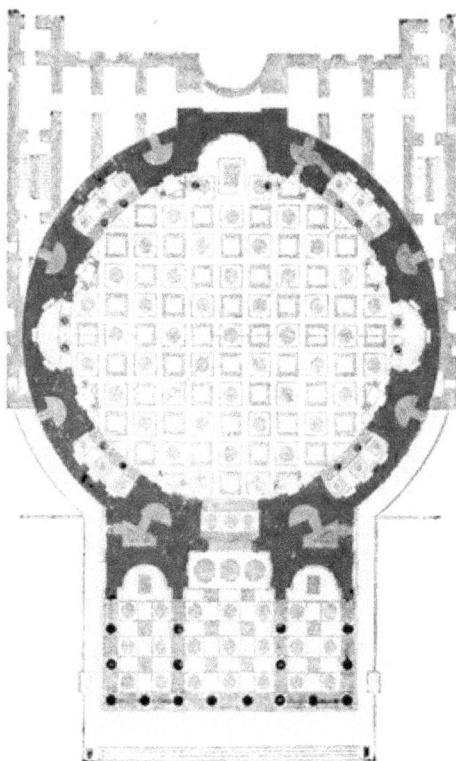

FIG. 15.—THE PANTHEON, ROME. GROUND-PLAN.

rotunda was most probably erected by Agrippa, the son-in-law of Augustus, in B.C. 27, and is a most remarkable instance of clever construction at so early a date. The

diameter of the interior is 145 ft. 6 in., and the height
to the top of the dome is 147 ft.   In addition to the er-
trance, the walls are broken up by seven large niches, three
of which are semicircular on plan, and the others, alter-
nating with them, rectangular.   The walls are divided

Fig. 126.—The Pantheon, Rome. Exterior.

into two stories by an entablature supported by columns
and pilasters; but although this is now cut through
by the arches of the niches, it is at least probable that
originally this was not the case, and that the entablature

ran continuously round the wall, as shown in Fig. 137,
which is a restoration of the Pantheon by Adler. Above
the attic story rises the huge hemispherical dome, which
is pierced at its summit by a circular opening 27 ft. in
diameter, through which a flood of light pours down and

FIG. 137.—THE PANTHEON, ROME. INTERIOR.

illuminates the whole of the interior. The dome is enriched by boldly recessed panels, and these were formerly
covered with bronze ornaments, which have been removed
for the sake of the metal. The marble enrichments of
the attic have also disappeared, and their place has been
taken by common and tawdry decorations more adapted
to the stage of a theatre. But notwithstanding everything that has been done to detract from the imposing
effect of the building by the alteration of its details, there
is still, taking it as a whole, a simple grandeur in the

FIG. 133.—THE CORINTHIAN ORDER FROM THE PANTHEON, ROME.

design, a magnificence in the material employed, and a quiet harmony in the illumination, that impart to the interior a character of sublimity which nothing can impair. The rectangular portico was added at some subsequent period, and consists of sixteen splendid Corinthian columns (Fig. 138), eight in front supporting the pediment, and the other eight dividing the portico into three bays, in precisely the same way as if it formed the pronaos to the three cells of an Etruscan temple.

Bridges and Aqueducts.

The earliest Roman bridges were of wood, and the Pons Sublicius, though often rebuilt, continued to be of this material until the time of Pliny, but it was impossible for a people who made such use of the arch to avoid seeing the great advantage this form gave them in the construction of bridges, and several of these formed of stone spanned the Tiber even before the time of the Empire. The finest Roman bridges, however, were built in the provinces. Trajan constructed one over the Danube which was 150 ft. high and 60 ft. wide, and the arches of which were of no less than 170 ft. span. This splendid structure was destroyed by his successor, Hadrian, who was probably jealous of it. The bridge over the Tagus at Alcantara, which was constructed by Hadrian, is another very fine example. There were six arches here, of which the two centre ones had a span of 100 ft.

The Roman aqueducts afford striking evidence of the building enterprise and architectural skill of the people. Pliny says of these works : "If any one will carefully consider the quantity of water used in the open air, in private baths, swimming-baths, houses, gardens, &c., and thinks

of the arches that have been built, the hills that have been
tunnelled, and the valleys that have been levelled for the
purpose of conducting the water to its destination, he must
confess that nothing has existed in the world more calculated to excite admiration." The same sentiment strikes
an observer of to-day when looking at the ruins of these
aqueducts. At the end of the first century A.D. we read of
nine aqueducts in Rome, and in the time of Procopius
(A.D. 550) there were fourteen in use. Of these, the Aqua
Claudia and the Anio Novus were the grandest and most
costly. These were constructed about the year 48 A.D.,
and entered the city upon the same arches, though at
different levels, the Aqua Claudia being the lower. The
arches carrying the streams were over nine miles long,
and in some cases 109 ft. high. They were purely works
of utility, and had no architectural decorations; but they
were most admirably adapted for their purpose, and were
so solidly constructed, that portions of them are still in
use. Some of the provincial aqueducts, such as those of
Tarragona and Segovia in Spain, were more ornamental,
and had a double tier of arches. The Pont du Gard, not
far from Nîmes, in France, is a well-known and very
picturesque structure of this character.

Commemorative Monuments.

These comprise triumphal arches, columns, and tombs.
The former consisted of a rectangular mass of masonry
having sculptured representations of the historical event
to be commemorated, enriched with attached columns on
pedestals, supporting an entablature crowned with a high
attic, on which there was generally an inscription. In
the centre was the wide and lofty arched opening. The

Arch of Titus, recording the capture of Jerusalem, is one
of the finest examples. Later on triumphal arches were
on a more extended scale, and comprised a small arch on
each side of the large one; examples of which may be

FIG. 139.—THE ARCH OF CONSTANTINE, ROME.

seen in the arches of Septimius Severus and of Constantine
(Fig. 139). The large arched gateways which are met
with in various parts of Europe—such as the Porte
d'Arroux at Autun, and the Porta Nigra at Trèves—are

monuments very similar to triumphal arches. There remain also smaller monuments of the same character, such as the so-called Arch of the Goldsmiths in Rome (Fig. 1).

Columns were erected in great numbers during the time of the Emperors as memorials of victory. Of these the Column of Trajan and that of Marcus Aurelius are the finest. The former was erected in the centre of Trajan's Forum, in commemoration of the Emperor's victory over the Dacians. It is of the Doric order, 132 ft. 10 in. high, including the statue. The shaft is constructed of thirty-four pieces of marble joined with bronze cramps. The figures on the pedestal are very finely carved, and the entire shaft is encircled by a series of elaborate bas-reliefs winding round it in a spiral from its base to its capital. The beauty of the work on this shaft may be best appreciated by a visit to the cast of it set up—in two heights, unfortunately—at the South Kensington Museum. The Column of Marcus Aurelius, generally known as the Antonine Column, is similarly enriched, but is not equal to the Trajan Column.

The survival of Etruscan habits is clearly seen in the construction of Roman tombs, which existed in enormous numbers outside the gates of the city. Merivale says: "The sepulchres of twenty generations lined the sides of the high-roads for several miles beyond the gates, and many had considerable architectural pretensions." That of Cecilia Metella is a typical example. Here we find a square basement surmounted by a circular tower-like structure, with a frieze and cornice. This was erected about B.C. 60, by Crassus. The mausoleum of Augustus was on a much more extensive scale, and consisted of four cylindrical stories, one above the other, decreasing in diameter as they ascended, and the topmost of all was

crowned with a colossal statue of the Emperor. The
tomb of Hadrian, on the banks of the Tiber—now known
as the Castle of Sant' Angelo—was even more magnificent.
This comprised a square base, 75 ft. high, the side of
which measured about 340 ft. ; above this was a cylindrical building surmounted by a circular peristyle of thirty-
four Corinthian columns. On the top was a quadriga
with a statue of the Emperor. These mausolea were occasionally octagonal or polygonal in plan, surmounted by
a dome, and cannot fail to remind us of the Etruscan
tumuli.

Another kind of tomb, of less magnificence, was the
columbarium, which was nothing more than a subterranean chamber, the walls of which had a number of
small apertures in them for receiving the cinerary urns
containing the ashes of the bodies which had been
cremated. In the eastern portion of the Empire, in rocky
districts, the tombs were cut in the rock, and the façade
was elaborately decorated with columns and other architectural features.

Domestic Architecture.

Of all the palaces which the Roman emperors built for
themselves, and which we know from historical records to
have been of the most magnificent description, nothing
now remains in Rome itself that is not too completely
ruined to enable any one to restore its plan with accuracy,
though considerable remains exist of the Palace of the
Cæsars on the Palatine Hill. In fact, the palace of
Diocletian at Spalatro, in Dalmatia, is the only remaining example in the whole of the Roman empire of the
dwelling-house of an emperor, and even this was not
built till after Diocletian had resigned the imperial dig-

nity, so that its date is the early part of the fourth century A.D. This palace is a rectangle, measuring about 700 ft. one way and 590 ft. the other, and covers an area of nearly 10 acres. It is surrounded by high walls, broken at intervals by square and octagonal towers, and contains temples, baths, and extensive galleries, besides the private apartments of the Emperor and dwellings for the principal officers of the household. The architect of this building broke away from classical traditions to a great extent; for example, the columns stand on corbels instead of pedestals, the entablatures being much broken, and the arches spring directly from the capitals of the columns (Fig. 149).

The private houses in Rome were of two kinds: the *insula* and the *domus*. The insula was a block of buildings several stories high, frequently let out to different families in flats. The ground-floor was generally given up to shops, which had no connection with the upper parts of the building; and one roof covered the whole. This kind of house was generally tenanted by the poorer class of tradesmen and artificers. The other kind of house, the domus, was a detached mansion. The excavations at Pompeii have done much to elucidate a number of points in connection with Roman dwellings which had been the subject of much discussion by scholars, but we must not too hastily assume that the Pompeian houses are the exact counterpart of those of ancient Rome, as Pompeii was what may be called a Romano-Greek city.

The general arrangements of a Roman house were as follows: next the street an open space was frequently left, with porticoes on each side of it provided with seats: this constituted the vestibule, and was entirely outside the house;* the entrance-door opened into a

---

* This does not occur in the Pompeian houses.

narrow passage, called the *prothyrum*, which led to the *atrium*,* which in the houses of Republican Rome

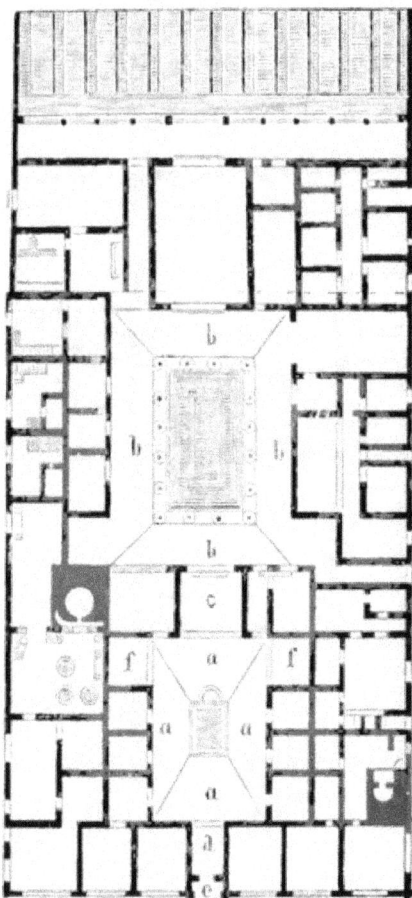

FIG. 119.—GROUND-PLAN OF THE HOUSE OF PANSA, POMPEII.

was the principal apartment, though afterwards it served as a sort of waiting-room for the clients and retainers

* Marked *a*, *a*, on the plans.

of the house; it was an open court, roofed in on all the four sides, but open to the sky in the centre. The simplest form was called the Tuscan atrium, where the roof was simply a lean-to sloping towards the centre, the rafters being supported on beams, two of which rested on the walls of the

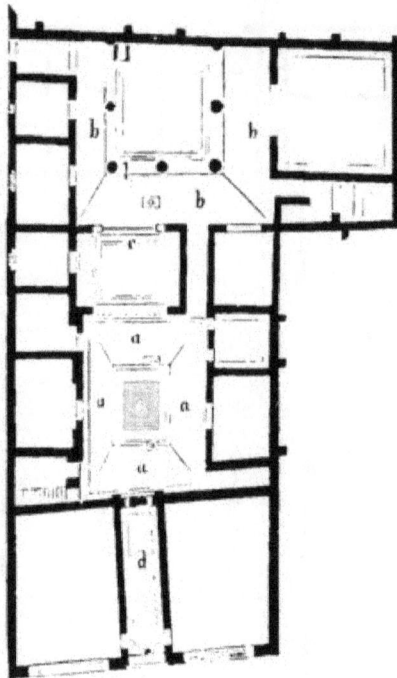

FIG. 141.—GROUND-PLAN OF THE HOUSE OF THE TRAGIC POET, POMPEII.

atrium, and had two other cross-beams trimmed into them. The centre opening was called the *impluvium*, and immediately under it a tank, called the *compluvium*, was formed in the pavement to collect the rain-water (Fig. 142). When the atrium became larger, and the roof had to be

N

supported by columns, it was called a *cavædium.**   At the
end of this apartment were three others, open in front, the

FIG. 112.—THE ATRIUM OF A POMPEIAN HOUSE.

largest, in the centre, called *tablinum,* and the two side ones

* Vitruvius, however, seems to use the terms *atrium* and *cavædium*
as quite synonymous.

alæ ; * these were muniment-rooms, where all the family archives were kept, and their position is midway between the semi-public part of the house, which lay towards the front, and the strictly domestic and private part, which lay in the rear. At the sides of the atrium in the larger houses were placed small rooms, which served as sleeping chambers.

From the end of the atrium a passage, or sometimes two passages, called the *fauces*, running by the side of the tablinum, led to the *peristylium*,† which was the grand private reception-room; this also was a court open to the sky in the centre, and among the wealthy Romans its roof was supported by columns of the rarest marbles. Round the peristyle were grouped the various private rooms, which varied according to the size of the house and the taste of the owner. There was always one dining-room (*triclinium*), and frequently two or more, which were arranged with different aspects, for use in different seasons of the year. If several dining-rooms existed, they were of various sizes and decorated with various degrees of magnificence; and a story is told of one of the most luxurious Romans of Cicero's time, that he had simply to tell his slaves which room he would dine in for them to know what kind of banquet he wished to be prepared. In the largest houses there were saloons (*œci*), parlours (*exedræ*), picture galleries (*pinacothecæ*), chapels (*lararia*), and various other apartments. The kitchen, with scullery and bakehouse attached, was generally placed in one angle of the peristyle, round which various sleeping-chambers, according to the size of the house, were arranged. Most of the rooms appear to have been on the ground-floor, and probably depended for their light upon the

* Marked respectively *c*, and *f, f*, on the plan of the House of Pansa.
† Marked *b, b*, on the plans.

doorway only ; though in some instances at Pompeii small
windows exist high up in the walls.

Fig. 113.—Wall Decoration from Pompeii.

In the extreme rear of the larger houses there was
generally a garden ; and in those which were without this,

the dead walls in the rear were frequently painted so as to imitate a garden. The houses of the wealthy Romans were decorated with the utmost magnificence: marble columns, mosaic pavements, and charming pieces of sculpture adorned their apartments, and the walls were in all cases richly painted (Fig. 143), being divided into panels, in the centre of which were represented sometimes human figures, sometimes landscapes, and sometimes pictures of historical events. All the decoration of Roman houses was internal only: the largest and most sumptuous mansion had little to distinguish it, next the street, from a comparatively humble abode; and, with the exception of the space required for the vestibule and entrance doorway, nearly the whole of the side of the house next the street was most frequently appropriated to shops. All that we are able to learn of the architecture of Roman private houses, whether from contemporary descriptions or from the uncovered remains of Pompeii and Herculaneum,* points to the fact that it, even in a greater measure than the public architecture, was in no sense of indigenous growth, but was simply a copy of Greek arrangement and Greek decoration.

* At the Crystal Palace can be seen an interesting reproduction of a Pompeian house, which was designed by the late Sir Digby Wyatt. It gives a very faithful reproduction of the arrangement and the size of an average Pompeian house; and though every part is rather more fully covered with decoration than was usual in the originals, the decorations of each room faithfully reproduce the treatment of some original in Pompeii or Herculaneum.

FIG. 141.—CARVING FROM THE FORUM OF NERVA, ROME.

CHAPTER X.

ROMAN ARCHITECTURE.

Analysis.

THE *Plan (or floor-disposition).*—The plans of Roman buildings are striking from their variety and the vast extent which in some cases they display, as well as from a certain freedom, mastery, and facility of handling which are not seen in earlier work. Their variety is partly due to the very various purposes which the buildings of the Romans were designed to serve: these comprised all to which Greek buildings had been appropriated, and many others, the product of the complex and luxurious civilisation of the Empire. But independent of this circumstance, the employment of such various forms in the plans of buildings as the ellipse, the circle, and the octagon, and their facile use, seem to denote a people who could build rapidly, and who looked carefully to the general masses and outlines of what they built, however carelessly they handled the minute details. The freedom with which these new forms were employed arises partly also from the fact that the Romans were in possession of a system of construction which rendered them practically independent of most of the restrictions which had fettered the genius of the Egyptians, Assyrians, and Greeks. Their vaulted roofs could be supported by a comparatively small number of piers of great solidity, placed far apart; and accordingly in the great halls of the Thermæ and elsewhere we find planning in which, a few stable points of support being secured, the outline of the spaces between them is varied at the pleasure of the architect in the most picturesque and pleasing manner.

The actual floor received a good deal of attention from the Romans. It was generally covered with tesselated pavement, often with mosaic, and its treatment entered into the scheme of the design for most interiors.

The Walls.

The construction of these was essentially different from that adopted by most earlier nations. The Romans rather avoided than cultivated the use of large blocks of stone; they invented methods by which very small materials could be aggregated together into massive and solid walls. They used mortar of great cementing power, so much so that many specimens of Roman walling exist in this country as well as in Italy or France, where the mortar is as hard as the stones which it unites. They also employed a system of binding together the small materials so employed by introducing, at short distances apart, courses of flat stones or bricks, called "bond courses," and they further fortified such walls by bands of flat materials placed edgeways after the manner popularly known as herring-bone work. The result of these methods of construction was that the Roman architect could build anywhere, no matter how unpromising the materials which the locality afforded; that he could put the walls of his building together without its being requisite to employ exclusively the skilled labour of the mason, and that both time and expense were thus saved. This economy and speed were not pushed so far as to render the work anything but durable; they had, however, a bad effect in another direction, for these rough rubble walls were habitually encased in some more sightly material, in order to make them look as though they were something finer than they really were; and accordingly, the exterior was often faced with a thin skin of masonry, and not infrequently plastered. The interior was also almost invariably plastered, but to this little exception can be taken. This casing of the exteriors was, however, the beginning of a system of what may be called false architecture, and one which led to much that was degrading to the art.

The walls were in many cases, it has been already observed, gathered into strong masses, such as it is customary to term piers, in order to support the vaulted roofs at the proper points. They were often carried to a much greater height than in Greek buildings, and they played altogether a far more important part in the design of Roman buildings than they had done in that of the Greeks.

The Roofs.

As has been already stated, the Romans, in their possession of a new system of construction, enjoyed a degree of freedom which was unknown before. This system was based upon the use of the arch, and arched roofs and domes, and it enabled the Romans to produce interiors

unapproached before for size and splendour, and such as
have hardly been surpassed since, except by the vaulted
churches of the Middle Ages,—buildings which are themselves descended from Roman originals. The art of
vaulting was. in short, the key to the whole system of
Roman architecture, just as the Orders were to that of
the Greeks.

The well-known arch over the Cloaca Maxima at Rome
(Fig. 123, p. 142) may be taken as an illustration of the
most ancient and most simple kind of vault, the one which
goes by the significant name of "barrel or waggon-head
vault." This is simply a continuous arched vault springing from the top of two parallel walls; in fact, like the
arch of a railway tunnel. Such a vault may be constructed of very great span, and affords a means of
putting a permanent roof over a floor the outline of which
is a parallelogram; but it is heavy and uninteresting in
appearance. It was soon found to be possible to introduce
a cross vault running at right angles to the original one ;
and where such an intersecting vault occurs the side walls
of the original vault may be dispensed with, for so much
of their length as the newly-added vault spans.

The next step was to introduce a succession of such cross
vaults close to one another, so that large portions of the
original main wall might be dispensed with. What remained of the side walls was now only a series of oblong
masses or piers, suitably fortified so as to carry the great
weight resting upon them, but leaving the architect free to
occupy the space between them as his fancy might dictate,
or to leave it quite open. In this way were constructed the
great halls of the Thermæ; and the finest halls of modern
classic architecture—such, for example, as the Madeleine
at Paris, or St. George's Hall at Liverpool—are only a

reproduction of the splendid structures which such a
system of vaulting rendered possible.

When the floor of the space to be vaulted was circular,
the result of covering it with an arched roof was the dome
—a familiar feature of Roman architecture, and the noblest
of all forms of roof. We possess in the dome of the
Pantheon a specimen, in fairly good preservation, of this
kind of roof on the grandest scale.

We shall find that in later ages the dome and the vault
were adopted by the Eastern and the Western schools of
Christian architecture respectively. In Rome we have the
origin of both.

The Openings.

These were both square-headed and arched; but the
arched ones occur far more frequently than the others, and,
when occasion required, could be far bolder. The openings became of much greater importance than in earlier
styles, and soon disputed with the columns the dignity of
being the feature of the building: this eventually led, as
will be related under the next head, to various devices for
the fusion of the two.

The adoption of the arch by the Romans led to a
great modification in classic architecture; for its influence was to be traced in every part of the structure
where an opening of any sort had to be spanned. Formerly the width of such openings was very limited,
owing to the difficulty of obtaining lintels of great length.
Now their width and height were pure matters of choice,
and doorways, windows, and arcades naturally became
very prominent, and were often very spacious.

The Columns.

These necessarily took an altered place as soon as buildings were carried to such a height that one order could not, as in Greek temples, occupy the whole space from pavement to roof. The Greek orders were modified by the Romans in order to fit these altered circumstances, but columnar construction was by no means disused when the arch came to play so important a part in building. The Roman Doric order, and a very simple variety of it called Tuscan, were but rarely used. The chief alteration from the Greek Doric, in addition to a general degradation of all the mouldings and proportions, was the addition of a base, which sometimes consists of a square plinth and large torus, sometimes is a slightly modified Attic base; the capital has a small moulding round the top of the abacus, and under the ovolo are two or three small fillets with a necking below; the shaft was from 6 to 7 diameters in height, and was not fluted; the frieze was ornamented with triglyphs, and the metopes between these were frequently enriched with sculptured heads of bulls: the metopes were exact squares, and the triglyphs at the angles of buildings were placed precisely over the centre of the column.

The Ionic order was but slightly modified by the Romans, the chief alteration being made in the capital. Instead of forming the angular volutes so that they exhibited a flat surface on the two opposite sides of the capital, the Romans appear to have desired to make the latter uniform on all the four sides; they therefore made the sides of the abacus concave on plan, and arranged the volutes so that they seemed to spring out of the mouldings under the abacus and faced anglewise. The capital altogether seems

compressed and crowded up, and by no means elegant;
in fact, both this and the Doric order were decidedly de-
teriorations from the fine forms of Greek architecture.

The Corinthian order was much more in accordance
with the later Roman taste for magnificence and display,
and hence we find its use very general both in Rome and
in other cities of the Empire. Its proportions did not
greatly differ from those of the Greek Corinthian, but
the mouldings in general were more elaborate. Numerous
variations of the capital exist (Figs. 145, 145*a*), but the
principal one was an amalgamation of the large Ionic
volutes in the upper with the acanthus leaves of the lower

FIG. 145.—ROMAN CORINTHIAN CAPITAL
AND BASE FROM THE TEMPLE OF VESTA
AT TIVOLI.

FIG. 145*a*.—THE ROMAN COMPOSITE CAPITAL.

portion of the capital: this is known as the Composite
order, and the capital thus treated has a strength and vigour
which was wanting to the Greek order (see Fig. 145*a*).

The shafts of the columns were more often fluted than
not, though sometimes the lower portion was left plain
and the upper only fluted. The Attic base was generally
used, but an example has been found of an adaptation of
the graceful Persepolitan base to the Corinthian column.
This was the happiest innovation that the Romans made;
it seems, however, to have been but an individual attempt,
and, as it was introduced very shortly before the fall of
the Empire, the idea was not worked out.

The orders thus changed were employed for the most
part as mere decorative additions to the walls. In many
cases they did not even carry the eaves to the roof, as
they always did in a Greek temple; and it was not
uncommon for two, three, or more orders to be used one
above another, marking the different stories of a lofty
building.

The columns, or pilasters which took their place, being
reduced to the humble function of ornaments added to the
wall of a building, it became very usual to combine them
with arched openings, and to put an arch in the interspace
between two columns, or, in other words, to add a column
to the pier between two arches (Fig. 146). These arched
openings being often wide, a good deal of disproportion
between the height of the columns and their distance
apart was liable to occur; and, partly to correct this, the
column was often mounted upon a pedestal, to which the
name of "stylobate" has been given.

It was also sometimes customary to place above the
order, or the highest order where more than one was
employed, what was termed an attic—a low story ornamented with piers or pilasters. The exterior of the
Colosseum (Fig. 5), the triumphal arches of Constantine
(Fig. 139) and Titus, and the fragments of the upper part

FIG. 140.—PART OF THE THEATRE OF MARCELLUS, ROME. SHOWING THE COMBINATION OF COLUMNS AND ARCHED OPENINGS.

of the Forum of Nerva (Fig. 147) may be consulted as
illustrations of the combination of an order and an arched
opening, and of the use of pedestals and attics.

FIG. 147.—FROM THE RUINS OF THE FORUM OF NERVA, ROME. SHOWING THE USE
OF AN ATTIC STORY. WITH PLAN.

Another peculiarity, of which we give an illustration from
the baths of Diocletian (Fig. 148), was the surmounting a
column or pilaster with a square pillar of stone, moulded
in the same way as an entablature, *i.e.* with the regular
division into architrave, frieze, and cornice. This was a

decided perversion of the
use of the order; it occurs
in examples of late date.
So also do various other
arrangements for making
an arch spring from the
capital of a column; one
of these, from the palace
of Diocletian at Spalatro,
we are able to illustrate
(Fig. 149).

In conclusion, it may be
worth while to say that
the Roman writers and
architects recognised five
orders : the Tuscan, Doric,
Ionic, Corinthian, and
Composite, the first and
last in this list being,
however, really only vari-

FIG. 148.—FROM THE BATHS OF DIOCLETIAN,
ROME. SHOWING A FRAGMENTARY ENTAB-
LATURE AT THE STARTING OF PART OF A
VAULT.

FIG. 149.—FROM THE PALACE OF DIOCLETIAN, SPALATRO. SHOWING AN ARCH SPRINGING
FROM A COLUMN.

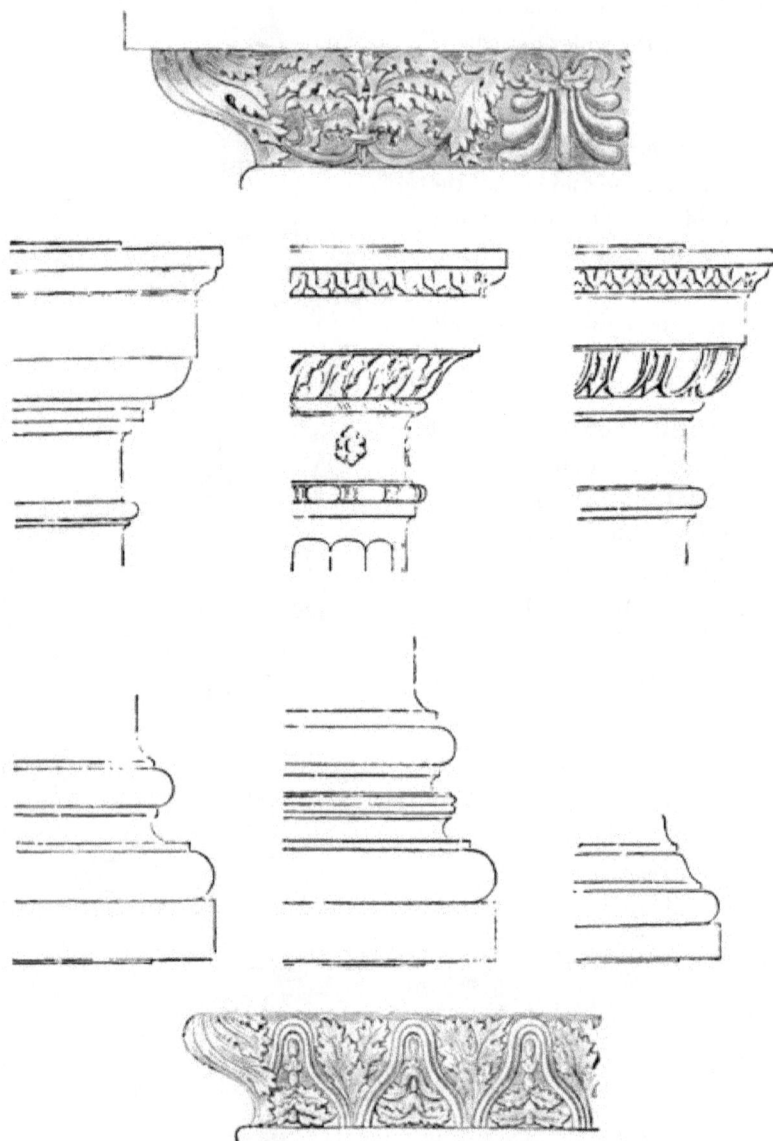

FIG. 150.—MOULDINGS AND ORNAMENTS FROM VARIOUS ROMAN BUILDINGS.

ations; and that when they placed the orders above one
another, they invariably used those of them which they
selected in the succession in which they have been named;
that is to say, the Tuscan or Doric lowest, and so on in
succession.

The Ornaments.

The mouldings with which Roman buildings are ornamented are all derived from Greek originals, but are often
extremely rough and coarse. It is true that in some old

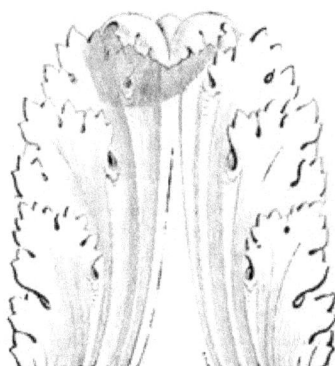

FIG. 151.—ROMAN CARVING. AN ACANTHUS LEAF

FIG. 152.—THE EGG AND DART ENRICHMENT. Roman.

Roman work, especially in those of the tombs which are
executed in marble, mouldings of considerable delicacy
and refinement of outline occur, but these are exceptional.
The profiles of the mouldings are, as a rule, segments of

circles, instead of being more subtle curves, and the result is that violent contrasts of light and shade are obtained, telling enough at a distance, but devoid of interest if the spectator come near.

Carving is executed exactly on the same principles as those which govern the mouldings—that is to say, with much more coarseness than in Greek work; not lacking

Fig. 153.—WALL DECORATION OF (SO-CALLED) ARABESQUE CHARACTER FROM POMPEII.

in vigour, or in a sort of ostentatious opulence of ornament, but often sadly deficient in refinement and grace.

Statues, many of them copies of Greek originals, generally executed with a heavy hand, but sometimes clearly of Greek work, were employed, as well as bronzes, inlaid marbles, mosaics, and various devices to ornament the interiors of Greco-Roman buildings; and free use was made of ornamental plaster-work, both on walls and vaults.

Coloured decoration was much in vogue, and, to judge

from what has come down to us, must have been executed
with great taste and much spirit. The walls of a Roman
dwelling-house of importance seem to have been all
painted, partly with that light kind of decoration to
which the somewhat inappropriate name of arabesque
has been given, and partly with groups or single figures,
relieved by dark or black backgrounds. The remains of
the Palace of the Cæsars in Rome, much of it not now
accessible, and the decorations visible at Pompeii, give a
high idea of the skill with which this mural ornamentation was executed; our illustration (Fig. 154) may be
taken as affording a good example of the combined decorations in relief and colour often applied to vaulted ceilings.

It is, however, characteristic of the lower level at
which Roman art stood as compared with Greek that,
though statues abounded, we find no traces of groups
of sculpture designed to occupy the pediments of
temples, or of bas-reliefs fitted to special localities in
the buildings, such as were all but universal in the best
Greek works.

Architectural Character.

The nature of this will have been to a large extent
gathered from the observations already made. Daring,
energy, readiness, structural skill, and a not too fastidious
taste were characteristic of the Roman architect and his
works. We find traces of vast spaces covered, bold construction successfully and solidly carried out, convenience
studied, and a great deal of magnificence attained in those
buildings the remains of which have come down to us;
but we do not discover refinement or elegance, a fine
feeling for proportion, or a close attention to details, to
a degree at all approaching the extent to which these
qualities are to be met with in Greek buildings. We

are thus sometimes tempted to regret that it was not possible to combine a higher degree of refinement with

FIG. 154.—DECORATION IN RELIEF AND COLOUR OF THE VAULT OF A TOMB IN THE VIA LATINA, NEAR ROME.

the great excellence in construction and contrivance exhibited by Roman architecture.

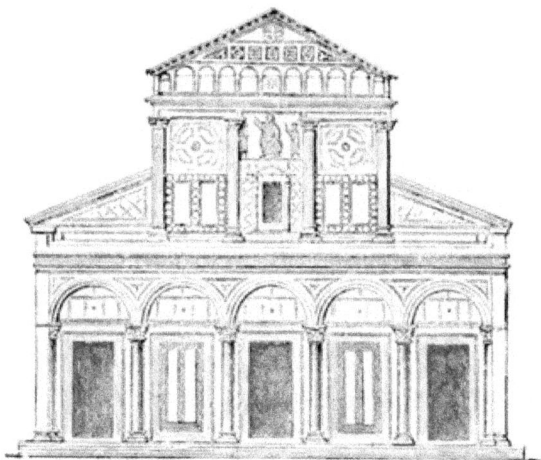

FIG. 55.—BASILICA-CHURCH OF SAN MINIATO, FLORENCE.

CHAPTER XI.

EARLY CHRISTIAN ARCHITECTURE.

Basilicas in Rome and Italy.

DURING the first three centuries the Christian religion was discredited and persecuted; and though many interesting memorials of this time (some of them having an indirect bearing upon architectural questions) remain in the Catacombs, it is chiefly for their paintings that the touching records of the past which have been preserved to us in these secluded excavations should be studied. Early in the fourth century Constantine the Great became Emperor, and in the course of his reign (from A.D. 312 to 337) he recognised Christianity, and made it the religion of the State. It then, of course, became requisite to provide places of public worship. Probably the Christians would have been, in many cases, reluctant to make use of heathen temples, and few temples, if any, were adapted to the assembling of a large congregation. But the large halls of the baths and the basilicas were free from associations of an objectionable character, and well fitted for large assemblages of worshippers. These and other such places were accordingly, in the first instance, employed as Christian churches. The basilica, however, became the model which, at least in Italy, was followed, to the exclusion of all others, when new buildings were erected for the purpose of Christian worship; and during the fourth century, and several succeeding ones, the churches of the West were all of the basilica type. What occurred at Constantinople, the seat of the Eastern Empire and the centre of the Eastern Church, will be considered presently.

There is probably no basilica actually standing which was built during the reign of Constantine, or near his time; but there are several basilica churches in Rome, such as that of San Clemente, which were founded near his time, and which, though they have been partially or wholly rebuilt, exhibit what is believed to be the ancient disposition without modification.

Access is obtained to San Clemente through a forecourt to which the name of the atrium is given. This is very much like the atrium of a Roman house, being covered with a shed roof round all four sides and open in the centre, and so resembling a cloister. The side next the church was called the narthex or porch; and when an atrium did not exist, a narthex at least was usually pro-

FIG 156.--INTERIOR OF A BASILICA AT POMPEII.

Restored, from descriptions by various authors.

vided. The basilica has always a central avenue, or nave,
and sides or aisles, and was generally entered from the
narthex by three doors, one to each division. The nave of
San Clemente is lofty, and covered by a simple wooden
roof; it is separated from the side aisles by arcades, the
arches of which spring from the capitals of columns; and
high up in its side walls we find windows. The side
aisles, like the nave, have wooden roofs. The nave termi-
nates in a semicircular recess called "the apse," the floor
of which is higher than that of the general structure, and
is approached by steps. A large arch divides this apse
from the nave. A portion of the nave floor is occupied by
an enclosed space for the choir, surrounded by marble
screens, and having a pulpit on either side of it. These
pulpits are termed "ambos." Below the Church of San
Clemente is a vaulted structure or crypt extending under
the greater part, but not the whole, of the floor of the
main building.

The description given above would apply, with very
slight variations, to any one of the many ancient basilica
churches in Rome, Milan, Ravenna, and the other older
cities of Italy; the principal variations being that in
many instances, including the very ancient basilica of
St. Peter, now destroyed, the avenues all stopped short of
the end wall of the basilica, and a wide and clear transverse space or transept ran athwart them in front of the
apse. San Clemente indeed shows some faint traces of
such a feature. In one or two very large churches five
avenues occur,—that is to say, a nave and double aisles;
and in Santa Agnese (Fig. 156A) and at least one other, we
find a gallery over the side aisles opening into the nave,
or, as Mr. Fergusson puts it, "the side aisles in two
stories." In many instances we should find no atrium,

Fig. 154.—Basilica or Early Christian Church of Santa Agnes at Rome

but in all cases we meet with the nave and aisles, and the apse at the end of the nave, with its arch and its elevated floor; and the entrances are always at the end of the building farthest from the apse, with some sort of porch or portal.

The interest of these buildings lies not so much in their venerable antiquity as in the fact that the arrangements of all Christian churches in Western Europe down to the Reformation, and of very many since, are directly derived from these originals. If the reader will refer to the description of a Gothic cathedral in the companion volume of this series,* it will not be difficult for him to trace the correspondence between its plan and its general structure and those of the primitive basilica. The atrium no longer forms the access to a cathedral, but it still survives in the cloister, though in a changed position. The narthex or porch is still more or less traceable in the great western portals, and in a kind of separation which often, but not always, exists between the westernmost bay of a cathedral. and the rest of the structure. The division into nave and aisles remains, and in very large churches and cathedrals there are double aisles, as there were in the largest basilicas. The nave roof is still higher than the aisles—the arcade, in two stories, survives in the usual arcade and triforium; the windows placed high in the nave are the present clerestory. The apsidal termination of the central avenue is still retained in almost all Continental architecture, though in Great Britain, from an early date, it was abandoned for a square east end; but square-ended or apsidal, a recess with a raised floor and a conspicuous arch, marking it off from the nave, always occupies this

* 'Gothic and Renaissance Architecture,' chap. ii. p. 6.

end of the church; and the under church, or crypt, is commonly, though not always, met with. The enclosure for the choir has, generally speaking, been moved farther east than it was in the basilica churches; though in Westminster Abbey, and in most Spanish cathedrals, we have examples of its occupying a position closely analogous to that of the corresponding enclosure at the Basilica of San Clemente. The cross passage to which we have referred as having existed in the old Basilica of St. Peter, and many others, is the original of the transept which in later churches has been made more conspicuous than it was in the basilica by being lengthened so as to project beyond the side walls of the church, and by being moved more westward. Lastly, the two ambos, or pulpits, survive in two senses. They are represented by the reading desk and the pulpit, and their situation and purpose are continued in the epistle and gospel sides of the choir.

The one point in which an essential difference occurs is the position of the altar, or communion table, and that of the Bishop's chair, or throne. In the classic basilica the apse was the tribunal, and a raised seat with a tesselated pavement occupied the central position in it, and was the justice-seat of the presiding judge; and in the sweep of the apse, seats right and left, at a lower elevation, were provided for assessors or assistant-judges. In front of the president was placed a small altar. The whole of these arrangements were copied in the basilica churches. The seat of the president became the bishop's throne, the seats for assessors were appropriated to the clergy, and the altar retained substantially its old position in front of the apse, generally with a canopy erected over it. This disposition continues in basilica churches to the present day.

At St. Peter's in Rome, for example, the Pope occupies a throne in the middle of the apse, and says mass with his face turned towards the congregation at the high altar,

FIG. 157.—SANT' APOLLINARE, RAVENNA. PART OF THE ARCADE AND APSE.

which stands in front of his throne under a vast baldacchino or canopy; but in Western Christendom generally a change has been made,—the altar has been placed in the

apse where the bishop's throne formerly stood, and the
throne of the bishop and stalls of his clergy have been
displaced, and are to be found at the sides of the choir or
presbytery.

Many basilica churches were erected out of fragments
taken from older buildings, and present a curious mix-
ture of columns, capitals, &c.; others, especially those at
Ravenna, exhibit more care, and are noble specimens of
ancient and severe architectural work. The illustration
which we give of part of the nave, arcade, and apse of one
of these, Sant' Apollinare in Classe, shows the dignified
yet ornate aspect of one of the most carefully executed of
these buildings (Fig. 157).

In some of these churches the decorations are chiefly
in mosaic, and are extremely striking. Our illustration of
the apse of the great basilica of St. Paul without the walls
(Fig. 158) may be taken as a fair specimen of the general
arrangement and treatment of the crowd of sacred figures
and subjects which it is customary to represent in these
situations; but it can of course convey no idea of the
brilliant effect produced by powerful colouring executed
in mosaic, the most luminous of all methods of enrichment. The floor of most of them was formed in the style
of mosaic known as "opus Alexandrinum," and the large
sweeping, curved bands of coloured material with which
the main outlines of the patterns are defined, and the
general harmony of colour among the porphyries and
other hard stones with which these pavements were executed, combine to satisfy the eye. A splendid specimen of
opus Alexandrinum, the finest north of the Alps, exists in
the presbytery of Westminster Abbey.

Another description of building is customarily met
with in connection with early Christian churches,—the

FIG. 158.—APSE OF THE BASILICA OF ST. PAUL WITHOUT THE WALLS, ROME.

baptistery. This is commonly a detached building, and almost always circular or polygonal. In some instances the baptistery adjoins the atrium or forecourt; but it soon became customary to erect detached baptisteries of considerable size. These generally have a high central portion carried by a ring of columns, and a low aisle running round, the receptacle for water being in the centre. The origin of these buildings is not so clear as that of the basilica churches; they bear some resemblance to the Roman circular temples; but it is more probable that the form was suggested by buildings similar in general arrangement, and forming part of a Roman bath. The octagonal building known as the baptistery of Constantine, and the circular building now used as a church and dedicated to Santa Costanza in Rome, and the celebrated baptistery of Ravenna, are early examples of this class of structure. Somewhat more recent, and very well known, are the great baptisteries of Florence and Pisa.

A few ancient circular or polygonal churches remain which do not appear to have been built as baptisteries. One of these is at Rome, the church of San Stefano Rotondo; but another, more remarkable in every way, is at Ravenna, the church of San Vitale. This is an octagonal building, with a large vestibule and a small apsidal choir. The central portion, carried by eight arches springing from as many lofty and solid piers, and surmounted by a hemispherical dome, rises high above the aisle which surrounds it. Much elegance is produced by the arrangement of smaller columns so as to form a kind of apsidal recess in each of the interspaces between the eight main piers.

Another feature which has become thoroughly identified with church architecture is the bell-tower, or campanile.

This appendage, there can be no doubt, originated with the basilicas of Italy. The use of bells as a call to prayer is said to have been introduced not later, at any rate, than the sixth century, and to this era is attributed a circular campanile belonging to Sant' Apollinare in Classe at Ravenna, a basilica already alluded to. The circular plan was, however, exceptional; the ancient campaniles remaining in Rome are all square; they are usually built of brick, many stories in height, and with a group of arched openings in each story, and are generally surmounted by a low conical roof.

The type of church which we have described influenced church architecture in Italy down to the eleventh century, and such buildings as the beautiful church (Fig. 155) of San Miniato, near Florence (A.D. 1013), and the renowned group of Cathedral, Baptistery, Campanile, and Campo Santo (a kind of cloistered cemetery) at Pisa, bear a very strong resemblance in many respects to those originals; though they belong rather to the Romanesque than to the Basilican division of early Christian architecture.

CHAPTER XII.

BYZANTINE ARCHITECTURE.

CONSTANTINE THE GREAT, who by establishing the Christian religion had encouraged the erection of basilicas for Christian worship in Rome and Italy, effected a great political change, and one destined to exert a marked influence upon Christian architecture, when he removed the seat of empire from Rome to Byzantium, and called the new capital Constantinople,* after his own name. Byzantium had been an ancient place, but was almost in ruins when Constantine, probably attracted by the unrivalled advantages of its site,† rebuilt it, or at least re-established it as a city. The solemn inauguration of Constantinople as the new capital took place A.D. 330 ; and when, under Theodosius, the empire was divided, this city became the capital of the East.

With a new point of departure among a people largely

---

* *I.e.* the City of Constantine.

† "The edge of the world : the knot which links together East and West ; the centre in which all extremes combine," was the not over-charged description given of Constantinople by one of her own bishops.

of Greek race, we might expect that a new development
of the church from some other type than the basilica
might be likely to show itself. This, in fact, is what
occurred; for while the most ancient churches of Rome
all present, as we have seen, an almost slavish copy of
an existing type of building, and do not attempt the use
of vaulted roofs, in Byzantium buildings of most original
design sprang up, founded, it is true, on Roman originals,
but by no means exact copies of them. In the erection
of these churches the most difficult problems of construction were successfully encountered and solved. What
may have been the course which architecture ran during
the two centuries between the refounding of Byzantium
and the building of Santa Sophia under Justinian, we can,
however, only infer from its outcome. It is doubtful if
any church older than the sixth century now remains in
Constantinople; but it is certain that, to attain the power
of designing and erecting so great a work as Santa Sophia,
the architects of Constantinople must have continued and
largely modified the Roman practice of building vaults and
domes. There is every probability that if some of the
early churches in Byzantium were domed structures others
may have been vaulted basilicas; the more so as the very
ancient churches in Syria, which owed their origin to
Byzantium rather than to Rome, are most of them of the
basilica type.

A church which had been erected by Constantine, dedicated to Santa Sophia (holy wisdom), was burnt early in
the reign of Justinian (A.D. 527 to 565); and in rebuilding
it his architects, Anthemios of Thralles, and Isodoros of
Miletus, succeeded in erecting one of the most famous
buildings of the world, and one which is the typical and
central embodiment of a distinct and very strongly marked

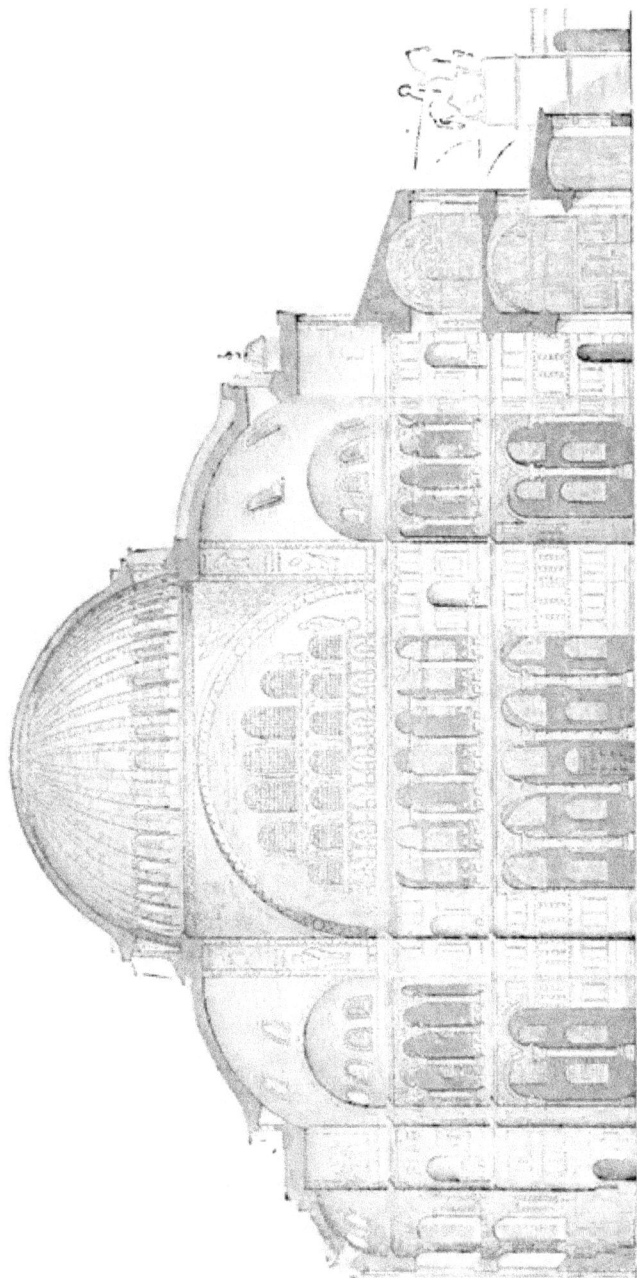

FIG. 149.—CHURCH OF SANTA SOPHIA AT CONSTANTINOPLE. LONG SECTION
BUILT UNDER JUSTINIAN BY ANTHEMIOS AND ISIDOROS. COMPLETED A.D. 537.

well-defined style. The basis of this style may be said to
be the adoption of the dome, in preference to the vault or
the timber roof, as the covering of the space enclosed within
the walls; with the result that the general disposition of
the plan is circular or square, rather than oblong, and
that the structure recalls the Pantheon more than the
great Hall of the Thermæ of Diocletian, or the Basilica of
St. Paul. In Santa Sophia one vast flattish dome domi-
nates the central space. This dome is circular in plan,
and the space over which it is placed is a square, the
sides of which are occupied by four massive semicircular
arches of 100 ft. span each, springing from four vast
piers, one at each of the four corners. The four triangular spaces between the corners of the square so
enclosed and the circle or ring resting upon it are filled
by what are termed "pendentives"—features which may,
perhaps, be best described as portions of a dome, each just
sufficient to fit into one corner of the square, and the
four uniting at their upper margin to form a ring. From
this ring springs the main dome. It rises to a height of
46 ft., and is 107 ft. in clear diameter. East and west of
the main dome are two half-domes, each springing from
a wall apsidal (i.e. semicircular) in plan. Smaller apses
again, domed over at a lower level, are introduced, and
vaulted aisles two stories in height occupy the sides of the
space within the outer walls till the outline of the building
is brought to very nearly an exact square. Externally this
church is uninteresting;* but its interior is of surpassing
beauty, and can be better described in the eloquent language of Gilbert Scott† than in any other: "Simple as

* For an illustration see Fig. 187.
† 'Lectures on Mediæval Architecture.'

is the primary ideal, the actual effect is one of great intricacy, and of continuous gradation of parts, from the small arcades up to the stupendous dome, which hangs with little apparent support like a vast bubble over the centre, or as Procopius, who witnessed its erection, described it, 'as if suspended by a chain from heaven.'

"The dome is lighted by forty small windows, which pierce it immediately above the cornice which crowns its pendentives, and which, by subdividing its lower part into narrow piers, increases the feeling of its being supported by its own buoyancy.

" The interior thus generated, covered almost wholly by domes, or portions of them, each rising in succession higher and higher towards the floating hemisphere in the centre, and so arranged that one shall open out the view to others, and that nearly the entire system of vaulting may be viewed at a single glance, appears to me to be in some respects the noblest which has ever been designed, as it was certainly the most daring which, up to that time at least, if not absolutely, had ever been constructed." After pointing out how the smaller arcades and apsidal projections, and the vistas obtained through the various arched openings, introduced intricate effects of perspective and constant changes of aspect, Scott continues : " This union of the more palpable with the more mysterious, of the vast unbroken expanse with the intricately broken perspective, must, as it appears to me, and as I judge from representations, produce an impression more astounding than that of almost any other building : but when we consider the whole as clothed with the richest beauties of surface,—its piers encrusted with inlaid marbles of every hue, its arcades of marble gorgeously carved, its domes and vaultings resplendent with gold mosaic interspersed

with solemn figures, and its wide-spreading floors rich
with marble tesselation, over which the buoyant dome
floats self-supported, and seems to sail over you as you
move,—I cannot conceive of anything more astonishing,
more solemn, and more magnificent."

The type of church of which this magnificent cathedral
was the great example has continued in Eastern Christendom to the present day, and has undergone surprisingly
little variation. A certain distinctive character in the
foliage (Fig. 163) employed in capitals and other decorative
carving, and mosaics of splendid colour but somewhat gaunt
and archaic design, though often solemn and dignified,
were typical of the work of Justinian's day, and could long
afterwards be recognized in Eastern Christian churches.

Between Rome and Constantinople, and well situated for
receiving influence from both those cities, stood Ravenna,
and here a series of buildings, all more or less Byzantine,
were erected. The most interesting of these is the church
of San Vitale (Figs. 160, 161). This building is octagonal
in plan, and thus belongs to the series of round and
polygonal churches and baptisteries for which the circular
buildings of the Romans furnished a model; but in its
high central dome, lighted by windows placed high up,
its many subsidiary arcades and apses, the latter roofed
by half-domes, and its vaulted aisles in two stories, it
recalls Santa Sophia; and its sculpture, carving, and
mosaic decorations are hardly less famous and no less
characteristic.

One magnificent specimen of Byzantine architecture,
more within the reach of ordinary travellers, and consequently better known than San Vitale or Santa Sophia,
must not be omitted, and can be studied easily by means
of numberless photographic illustrations—St. Mark's at

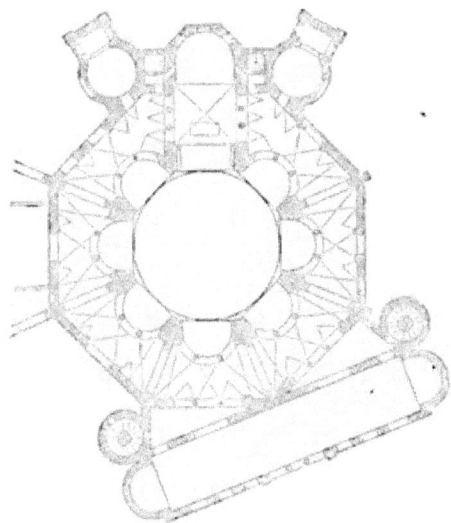

FIG. 160.—PLAN OF SAN VITALE AT RAVENNA.

FIG. 161.—SAN VITALE AT RAVENNA. LONGITUDINAL SECTION

Venice. This cathedral was built between the years
977-1071, and, it is said, according to a design obtained
from Constantinople. It has since been altered in external appearance by the erection of bulbous domical roofs
over its domes, and by additions of florid Gothic character;
but, disregarding these, we have alike in plan, structure,
and ornament, a Byzantine church of the first class.

The ground-plan of St. Mark's (Fig. 162) presents a
Greek cross, *i.e.* one in which all the arms are equal, and

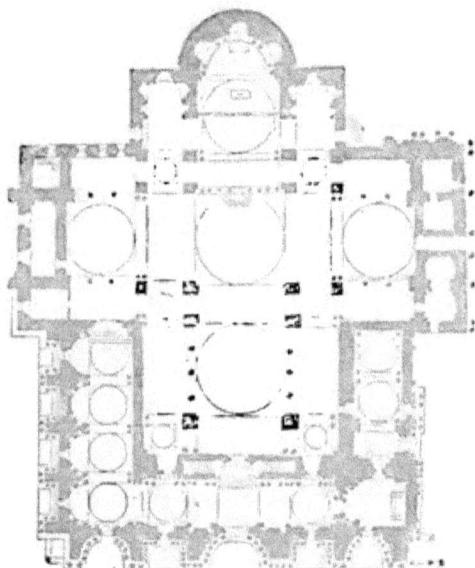

F.G. 162.—PLAN OF ST. MARK'S AT VENICE.

it is roofed by five principal domes, one at the crossing
and one over each of the four limbs of the cross. Aisles at
a low level, and covered by a series of small flat domes, in
lieu of vaulting, fill up the angles between the arms of the
cross, so as to make the outline of the plan nearly square.

The rich colouring of St. Mark's, due to a profuse employment of mosaics and of the most costly marbles, and the splendid effects produced by the mode of introducing light, which is admitted much as at Santa Sophia, are perhaps its greatest charm; but there is beauty in every aspect of its interior which has furnished a fit theme for the pen of the most eloquent writer on art and architecture of the present or perhaps of any day.

From Venice the influence of Byzantine art spread to a small extent in North Italy; in that city herself as well as in neighbouring towns, such as Padua, buildings and fragments of buildings exhibiting the characteristics of the style can be found. Remarkable traces of the influence of Byzantium as a centre, believed to be due to intercourse with Venice, can also be found in France. Direct communication with Constantinople by way of the Mediterranean has also introduced Byzantine taste into Sicily. One famous French church, St. Front in Périgueux, is identical (or nearly so) with St. Mark's in its plan; but all its constructive arches being pointed (Fig. 3, page 5), its general appearance differs a good deal from that of Eastern churches—a difference which is accentuated by the absence of the mosaics and other coloured ornaments which enrich the walls of St. Mark's. Many very old domed churches and much sculpture of the Byzantine type are moreover to be found in Central and Southern France—Anjou, Aquitaine, and Auvergne. These are, however, isolated examples of the style having taken root in spite of adverse circumstances; it is in those parts of Europe where the Greek Church prevails, or did prevail, that Byzantine architecture chiefly flourishes. In Greece and Asia Minor many ancient churches of Byzantine structure remain, while in Russia churches are built to the

present day corresponding to the general type of those
which have just been described.

In ancient buildings of Syria the influence of both the

FIG. 163.—FROM THE GOLDEN DOOR OF JERUSALEM. TIME OF JUSTINIAN. A.D. 500.

Roman and the Byzantine models can be traced. No
more characteristic specimens of Byzantine foliage can
be desired than some to be found in Palestine, as for ex-

ample the Golden Gate at Jerusalem, which we illustrate
(Fig. 163); but in the deserted cities of Central Syria a

Fig. 164.—Church at Tramanin in Syria.   4th and 5th century.

group of exceptional and most interesting buildings, both
secular and sacred, exists, which, as described by De Vogüé,*

* 'Syrie Centrale.'

seem to display a free and very original treatment based upon Roman more than Byzantine ideas. We illustrate the exterior of one of these, the church at Turmanin (Fig. 164). This is a building divided into a nave and aisles and with a vestibule. Two low towers flank the central gable, and it will be noticed that openings of depressed proportion, mostly semicircular-headed, and with the arches usually springing from square piers, mark the building; while the use made of columns strongly resembles the manner in which in later times they were introduced by the Gothic architects.

FIG. 165.—TOWER OF A RUSSIAN CHURCH.

CHAPTER XIII.

ROMANESQUE ARCHITECTURE.

THE term Romanesque is here used to indicate a
style of Christian architecture, founded on Roman
art, which prevailed throughout Western Europe from the
close of the period of basilican architecture to the rise of
Gothic; except in those isolated districts where the influence of Byzantium is visible. By some writers the signifi-
cance of the word is restricted within narrower limits;
but excellent authorities can be adduced for the employment of it in the wide sense here indicated. Indeed some
difficulty exists in deciding what shall and what shall not
be termed Romanesque, if any mo e restricted definition
of its meaning is adopted; while under this general term,
if applied broadly, many closely allied local varieties—as,
for example, Lombard, Rhenish, Romance, Saxon, and
Norman—can be conveniently included.

The spectacle which Europe presented after the re-
moval of the seat of empire to Byzantium and the in-
cursions of the Northern tribes was melancholy in the
extreme. Nothing but the church retained any semblance
of organised existence; and when at last some kind of
order began to emerge from a chaos of universal ruin,
and churches and monastic buildings began to be built in
Western Europe, all of them looked to Rome, and not
to Constantinople, as their common ecclesiastical centre.
It is not surprising that, as soon as differences between
the ritual of the Eastern and the Western Church sprang
up, a contrast between Eastern and Western architecture
should establish itself, and that the early structures of

the many countries where the Roman Church flourished never wandered far from the Roman type, with the exception of localities where circumstances favoured direct intercourse with the East. The architecture of the Eastern Church, on the other hand, adhered quite as closely to the models of Byzantium.

The style, so far as is known, was for a long time almost, if not absolutely, the same over a very large part of Western Christendom, and it has received from Mr. Freeman the appropriate designation of Primitive Romanesque. It was not till the tenth century, or later, that distinctive varieties began to make their appearance; and though that which was built earlier than that date has, through rebuildings and enlargements as well as natural decay, been in many cases swept away, still enough may be met

FIG. 106.—TOWER OF EARL'S BARTON CHURCH.

with to show us what the buildings of that remote time were like.

The churches are usually small, and have an apsidal east end. The openings are rude, with round-headed arches and small single or two-light windows, and the outer

walls are generally marked by flat pilasters of very
slight projection. Towers are common, and the openings
in them are often divided into two or more lights by
rude columns. The plan of these churches was founded
on the basilica type, but they do not exhibit the same
internal arrangement; and it is very noteworthy that
many of them show marks of having been vaulted, or at
least partly vaulted ; and not covered, as the basilicas
usually were, by timber roofs. Even a country so remote
as Great Britain possessed in the 10th century many buildings of Primitive Romanesque character ; and in such Saxon
churches as those of Worth, Brixworth, Dover, or Bradford,
and such towers as those of Earl's Barton (Fig. 166),
Trinity Church Colchester, Barnack, or Sompting, we have
specimens of the style remaining to the present day.

By degrees, as buildings of greater extent and more
ornament were erected, the local varieties to which reference has been made began to develop themselves.
In Lombardy and North Italy, for example, a Lombard
Romanesque style can be recognised distinctly ; here a
series of churches were built, many of them vaulted, but
not many of the largest size. Most of them were on
substantially the same plan as the basilicas, though a considerable number of circular or polygonal churches were
also built. Sant' Ambrogio at Milan, and some of the
churches at Brescia, Pavia, and Lucca, may be cited as
well-known examples of early date, and a little later the
cathedrals of Parma, Modena, and Piacenza (Fig. 167),
and San Zenone at Verona. These churches are all distinguished by the free use of small ornamental arches and
narrow pilaster-strips externally, and the employment of
piers with half-shafts attached to them, rather than
columns, in the arcades ; they have fine bell-towers ; cir-

cular windows often occupy the gables, and very frequently the walls have been built of, or ornamented with, coloured materials. The sculpture—grotesque, vigorous, and full of rich variety—which distinguishes many of these buildings, and which is to be found specially enriching the doorways, is of great interest, and began early to develop a character that is quite distinctive.

FIG. 167.—CATHEDRAL AT PIACENZA.

Turning to Germany, we find that a very strong resemblance existed between the Romanesque churches of that country and those of North Italy. At Aix-la-Chapelle a polygonal church exists, built by Charlemagne, and which tradition asserts was designed on the model of San Vitale at Ravenna. The resemblance is undoubted, but the German church is by no means an exact copy of Justinian's building. Early examples of German Romanesque exist in the cathedrals of Mayence, Worms, and Spires, and a steady advance was made till a point

Q

was reached (in the twelfth century) at which the style
may be said to have attained the highest development
which Romanesque architecture received in any country
of Europe.

The arcaded ornament (the arches being very frequently
open so as to form a real arcade) which was noticed as
occurring in Lombard churches, belongs also to German
ones, though the secondary internal arcade (triforium) is
absent from some of the early examples. Piers are used
more frequently than columns in the interiors, and are
often very plain. From an early date the use of a western
as well as an eastern apse seems to have been common
in Germany, and high western façades extending between
two towers were features specially met with in that
country. For a notice and some illustrations of the
latest and best phase of German Romanesque, which may
with propriety be termed "round-arched Gothic," the
reader is referred to the companion volume of this series.*

France exhibits more than one variety of Romanesque ;
for not only, as remarked in the chapter on Byzantine Art,
is the influence of Greek or Venetian artists traceable in
the buildings of certain districts, especially Périgueux, but
it is clear that in others the existence of fine examples
of Roman architecture (Fig. 168) affected the design of
buildings down to and during the eleventh century. This
influence may, for example, be detected in the use, in the
churches at Autun, Valence, and Avignon, of capitals,
pilasters, and other features closely resembling classic
originals, and in the employment through a great part of
Central and Northern France of vaulted roofs.

A specially French feature is the chevet, a group of

* 'Gothic and Renaissance Architecture,' chap. vii.

apsidal chapels which were built round the apse itself, and which combined with it to make of the east end of a great cathedral a singularly rich and ornate composition.

FIG. 168.—VAULTS OF THE EXCAVATED ROMAN BATHS, IN THE
MUSÉE DE CLUNY, PARIS.

This feature, originating in Romanesque churches, was retained in France through the whole of the Gothic period, and a good example of it may be seen in the large Romanesque church of St. Sernin at Toulouse, which we illustrate (Fig. 169). The transepts were usually well

Q 2

FIG. 163.—CHURCH OF ST. SERNIN, TOULOUSE.

marked. The nave arcades generally sprang from piers (Fig. 170), more rarely from columns. Arches are constantly met with recessed, *i.e.* in receding planes,* the first stage of progress towards a Gothic treatment, and are occasionally slightly moulded (Fig. 171). Western doorways are often highly enriched with sculpture; and the carving and sculpture generally, though often rude, are full of vitality. Towers occur, usually square, more rarely octagonal. Window-lights are frequently grouped two or more under one arch. Capitals of a basket-shape, and with a square abacus, often richly sculptured, are employed.

In Normandy, and generally in the North of France, round-arched architecture was excellently carried out, and churches remarkable both for their extent and their great dignity and solidity were erected. Generally speaking, however, Norman architecture, especially as met with in Normandy itself, is less ornate than the

FIG. 170.—NAVE ARCADE AT ST. SERNIN, TOULOUSE.

* 'Gothic and Renaissance Architecture,' chap. v. p. 62.

Romanesque of Southern France; in fact some of the best examples seem to suffer from a deficiency of ornament.

FIG. 171.—ARCHES IN RECEDING PLANES AT ST. SERNIN, TOULOUSE.

The large and well-known churches at Caen, St. Etienne, otherwise the Abbaye aux Hommes—interesting to Englishmen as having been founded by William the Conqueror immediately after the Conquest—and the Trinité, or Abbaye aux Dames, are excellent examples of early Norman architecture, but the student must not forget that additions have been made to them, which, if they add to their beauty, at the same time alter their character. For example, in St. Etienne, the upper part of the western towers and the fine spires with which they are crowned were built subsequent to the original structure, as was also, in all probability, the chevet, or eastern limb. It seems probable also that the vaulting may not be what was contemplated in the original plan.

St. Etienne is 364 ft. long, and is lofty in its proportions. It has a nave and aisles, arcades resting on piers, and strongly-marked transepts, and has two western towers with the gable of the nave between them. The west front is well designed in three stories, having strongly-marked vertical divisions in the buttresses of the towers, and equally distinct horizontal divisions in the three doorways below, and two ranges of windows, each of five lights, above. There is no circular west window. The nave and aisles are vaulted.

Besides other cathedral churches, such for example as those of Bayeux and Evreux, in which considerable parts of the original structures remain, there exist throughout Normandy and Brittany many parochial churches and monastic buildings, exhibiting, at least in some portions of their structure, the same characteristics as those of St. Etienne; and it is clear that an immense number of buildings, the beauty and even refinement of which are conspicuous, must have been erected in Northern France during the eleventh and the early years of the twelfth centuries, the period to which Norman architecture in France may be said to belong.

In Great Britain, as has been already pointed out, enough traces of Saxon—that is to say, Primitive Romanesque—architecture remain to show that many simple, though comparatively rude, buildings must have been erected previous to the Norman Conquest. Traces exist also of an influence which the rapid advance that had been made by the art of building as practised in Normandy was exerting in our island. The buildings at Westminster Abbey raised by Edward the Confessor, though they have been almost all rebuilt, have left just sufficient traces behind to enable us to recognise that they were of bold design. The plan of the Confessor's church was laid out upon a scale almost as large as that of the present structure. The monastic buildings were extensive. The details of the work were, some of them, refined and delicate, and resembled closely those employed in Norman buildings at that time. Thus it appears that, even had the Conquest not taken place, no small influence would have been exerted upon buildings in England by the advance then being made in France; but instead of a gradual improvement being so produced, a sudden and rapid revolution was effected by the complete conquest of the country and its occupation by nobles and ecclesiastics from Normandy, who, enriched by the plunder of the conquered country, were eager to establish themselves in permanent buildings.

Shortly after the Conquest distinctive features began to show themselves. Norman architecture in England soon became essentially different from what it was in Normandy, and we possess in this country a large series of fine works showing the growth of this imported style, from the early simplicity of the chapel in the Tower of London to such elaboration as that of the later parts of Durham Cathedral.

The number of churches founded or rebuilt soon after

the Norman Conquest must have been enormous, for in examining churches of every date and in every part of England it is common to find some fragment of Norman work remaining from a former church: this is very frequently a doorway left standing or built into walls of later date; and, in addition to these fragments, no small number of churches, and more than one cathedral, together with numerous castles, remain in whole or in part as they were erected by the original builders.

Norman architecture is considered to have prevailed in England for more than a century; that is to say, from the Conquest (1066) to the accession of Richard I. (1189). For some details of the marks by which Norman work can be recognised the reader is referred to the companion volume;* we propose here to give an account of the broader characteristics of the buildings erected during the prevalence of the style.

The oldest remaining parts of Canterbury Cathedral are specimens of Norman architecture executed in England immediately after the Conquest. This great church was rebuilt by Archbishop Lanfranc (whose episcopate lasted from 1070 to 1089), and in extent as laid out by him was very nearly identical with the existing structure; almost every portion has, however, been rebuilt, so that of his work only the towers forming transepts to the choir, and some other fragments, now remain. More complete and equally ancient is the chapel in the Tower of London, which consists of a small apsidal church with nave and aisles, vaulted throughout, and in excellent preservation. This building, though very charming, is almost destitute

---

* 'Gothic and Renaissance Architecture,' chap. ii. p. 23.

of ornament. A little more ornate, and still a good example of early Norman, is St. Peter's Church, Northampton (Fig. 172), the interior of which we illustrate. To these examples of early Norman we may add a large part of Rochester Cathedral, and the transepts of Winchester.

FIG. 172.—NORMAN ARCHES IN ST. PETER'S CHURCH, NORTHAMPTON.

The transepts of Exeter present a specimen of rather more advanced Norman work; and in the cathedrals of Peterborough and Durham the style can be seen at its best.

In most Norman buildings we find very excellent masonry and massive construction. The exteriors of west

fronts, transepts, and towers show great skill and care in
their composition, the openings being always well grouped,
and contrasted with plain wall-spaces; and a keen sense
of proportion is perceptible. The Norman architects had
at command a rich, if perhaps a rather rude, ornamentation, which they generally confined to individual features,
especially doorways; on these they lavished mouldings
and sculpture, the elaboration of which was set off by the
plainness of the general structure. In the interior of
the churches we usually meet with piers of massive proportion, sometimes round, sometimes octagonal, sometimes
rectangular, and a shaft is sometimes carried up the face
of the piers; as, for example, in Peterborough Cathedral
(Fig. 173). The capitals of the columns and piers have a
square abacus, and, generally speaking, are of the cushion-
shaped sort, commonly known as basket-capitals, and are
profusely carved. The larger churches have the nave
roofed with a timber roof, and at Peterborough there
is a wooden ceiling; in these cases the aisles only are
vaulted, but in some small churches the whole building has
been so covered. Buttresses are seldom required, owing
to the great mass of the walls; when employed they have
a very slight projection, but the same strips or pilasters
which are used in German Romanesque occur here also.
Low towers were common, and have been not unfrequently
preserved in cases where the rest of the building has been
removed. As the style advanced, the proportions of arcades
became more lofty, and shafts became more slender, decorative arcades (Fig. 174) became more common, and in
these and many other changes the approaching transition
to Gothic may be easily detected.

We have already alluded to the many Norman doorways
remaining in parish churches of which all other parts
have been rebuilt. These doorways are generally very

rich; they possess a series of mouldings sometimes springing from shafts, sometimes running not only round the arched head, but also up the jambs of the opening; and each moulding *is* richly carved, very often with a repetition of the same ornament on each voussoir of the arch. Occasionally, but not frequently, large portions of wall-surface are covered by a diaper; that is to say, an ornament constantly repeated so as to produce a general sense of enrichment.

Norman castles, as well as churches, were built in great numbers shortly

FIG. 113.—NAVE ARCADE, PETERBOROUGH CATHEDRAL.

after the Conquest, and not a few remain. The stronghold which a follower of the Conqueror built in order to establish himself on the lands granted him was always a very sturdy massive square tower, low in proportion

FIG. 174.—DECORATIVE ARCADE FROM CANTERBURY CATHEDRAL.

to its width, built very strongly, and with every provision for sustaining an attack or even a siege. Such a tower is called "a keep;" and in many famous castles, as for example the Tower of London, the keep forms the nucleus round which buildings and courtyards of later date have clustered. In some few instances, however, as for example at Colchester, the keep is the only part now standing, and it is probable that when originally built these Norman castles were not much encumbered with out-buildings. Rochester Castle is a fine example of a Norman keep, though it has suffered much from decay and injury.

FIG. 175.—HEDINGHAM CASTLE.

The very large Norman keep of the Tower of London, known as the White Tower, and containing the chapel already described, has been much modernised and altered, but retains the fine mass of its original construction.

Perhaps the best (and best-preserved) example is Hedingham Castle in Essex, which we illustrate (Figs. 175 and 176). From the remains of this building some idea

FIG. 176.—INTERIOR OF HEDINGHAM CASTLE.

of the interior of the hall—the chief room within a Norman keep—may be obtained, as well as of the general external appearance of such a structure.

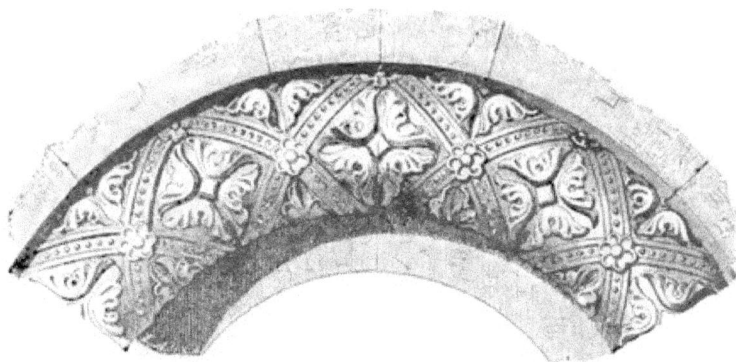

FIG. 177.—ROUNDED ARCH OF CHURCH AT GELNHAUSEN.

CHAPTER XIV.

CHRISTIAN ROUND-ARCHED ARCHITECTURE.

Analysis.

NOTWITHSTANDING very wide differences which undoubtedly exist, there is a sufficient bond of union between the Basilican, the Byzantine, and the Romanesque styles, to render it possible for us to include the characteristics of the three in an analysis of Christian round-arched architecture.

The Plan or floor-disposition of the basilican churches, as has been pointed out, was distinctive. The atrium, or forecourt, the porch, the division into nave and aisles; the transept, the great arch, and the apse beyond it with the episcopal seat at the back behind the altar; the ambos; and the enclosure for the choir, were typical features. Detached towers sometimes occurred. The plan of Romanesque churches was based upon that of the basilica; the atrium was often omitted, so was the transept sometimes; but, when retained, the transept was generally made more prominent than in the basilica. The position of the altar and of the enclosure for the choir were changed,

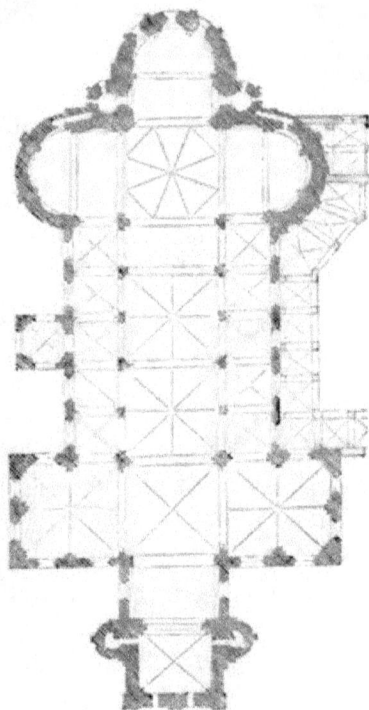

Fig. 178.—Plan of the Church of the Apostles at Cologne.

but in other respects the basilica plan was continued. In Germany, however, apsidal transepts (Fig. 178) were built. Towers were common, occasionally detached, but more frequently joined to the main building.

Circular and polygonal buildings for use as baptisteries,

R

and sometimes as churches, existed both in the basilican and the Romanesque time.

Byzantine church plans are all distinguished by their great central square space, covered by the central dome, flanked usually by four arms, comparatively short, and all of equal length; and the plan of the buildings is generally square, or nearly so, in outline. Circular and polygonal buildings sometimes occur.

Few traces of the arrangement of military, secular, or domestic buildings earlier than the twelfth century remain, but some examples of a cloister at the side of the nave (generally the south side) of a church, giving or intended to give access to monastic buildings, still exist.

FIG. 179.—SPIRE OF SPIRES CATHEDRAL.

The Walls of such buildings as have come down to us are, it may be well understood, strong, since the most recent of this round-arched series of buildings must be about seven hundred years old. Fine masonry was not much employed till the time of the Normans, but the Roman plan of building with bricks or rubble and casing the face of the walls with marble or mosaic, or at least plaster, was generally followed. The walls are carried up as gables and towers to a considerable extent (Fig. 179), especially in Western countries.

The Roof.—In a basilica this was of timber, in a Byzantine church it consisted of a series of domes ; in a Romanesque church it was sometimes of timber as in the basilica, but not unfrequently vaulted. As a general rule the vault prevailed in the West and the dome in the East ; and such examples of either sort of roof as occur in those provinces where the other was usual, like the domed churches in parts of France, must be looked upon as exceptional.

The Openings are almost invariably arched, and seldom, if ever, covered by a lintel. It is hardly necessary to add that the arches are always round. Almost always they are semicircular, but instances of the employment of a segmental arch, or of one the outline of which is a little more than half a circle, may be occasionally met with.

Door openings are often made important both by size and decoration. Window openings are usually small ; and the grouping of two or more lights under one head, which was so conspicuous a feature in Gothic architecture, first appears in Byzantine buildings, and is met with also in Romanesque ones. The mode of introducing light is to a certain extent characteristic. The basilican churches always possess a clerestory, and usually side windows in the aisles ; and this arrangement is generally followed in Romanesque buildings, though sometimes, in Germany, the clerestory is omitted. The gable ends of the nave and transepts are not usually pierced by many or large lights (Fig. 180) ; and when there is a central feature, as a tower, or even a dome, little or no light is introduced through it. On the other hand, the Byzantine churches depend largely for light upon the ring of windows which commonly encircles the base of the central dome, and sometimes that of the subsidiary domes ; and the gables are

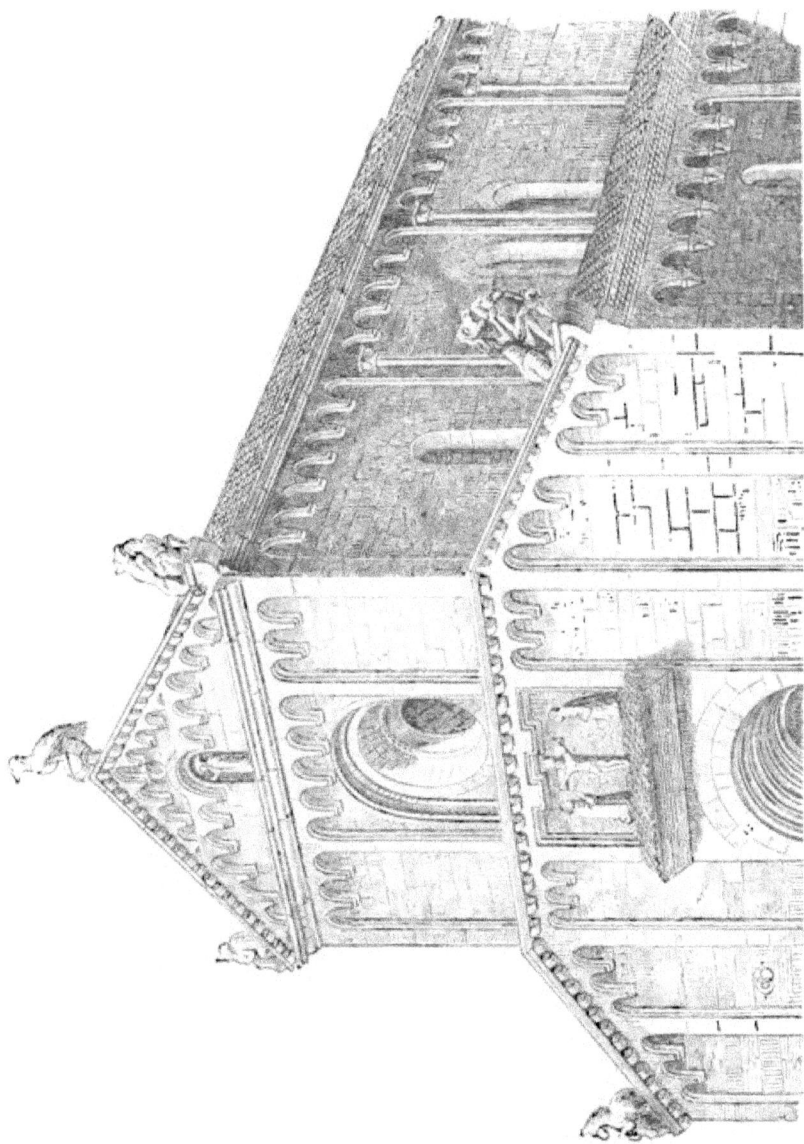

pierced so as to supply any additional light required, so
that windows are infrequent in the lower walls. Broadly
speaking, therefore, the Western churches have side-
lighting and the Eastern top-lighting.

The great arches which carry the main domes form a
notable feature in Eastern churches, and are of very bold
construction. In the basilican churches one great arch,
called " the arch of triumph," occurs, and only one ; this
gives access to the apse : and a similar arch, which we
now denominate " the chancel arch," usually occupies a
corresponding position in all Romanesque churches. The
arches of the arcade separating the nave from the aisles
in all Western churches are usually of moderate span.
In some ancient basilicas these arches are replaced by
a horizontal beam.

The Columns.—In basilicas these were of antique type ;
very often they had actually been obtained by the demo-
lition of older buildings, and when made purposely they
were as a rule of the same general character. The same
might be said of those introduced into Byzantine buildings, though a divergence from the classic type soon
manifested itself, and small columns began to appear as
decorative features. In Romanesque buildings the columns
are very varied indeed, and shafts are frequently introduced into the decoration of other features. They occur
in the jambs of doorways with mouldings or sub-arches
springing from them ; long shafts and short ones, frequently supporting ornamental arcades, are employed
both internally and externally ; and altogether that use
of the column as a means of decoration, of which Gothic
architecture presents so many examples, first began in the
Romanesque style.

The capitals employed in Romanesque buildings generally depart considerably from the classic type, being based on the primitive cube capital (Fig. 181), but, as a rule, in Eastern as well as in basilican churches, they bear a tolerably close resemblance to classic ones.

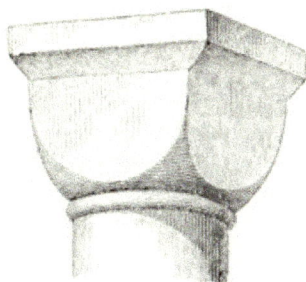

FIG. 181.—CUBIC CAPITAL.

The Ornaments throughout the whole of the Christian round-arched period are a very interesting subject of study, and will repay close attention. In the basilican style mouldings occur but seldom : where met with, they are all of the profiles common in Roman architecture, but often rudely and clumsily worked. Carving partakes also of classic character, though it is not difficult to detect the commencement of that metamorphosis which was effected in Byzantium, and which can hardly be better described than in the following paragraph from the pen of Sir Digby Wyatt:—"The foliage is founded on ancient Greek rather than on Roman traditions, and is characterised by a peculiarly sharp outline. All ornamental sculpture is in comparatively low relief, and the absence of human and other figures is very marked. Enrichments were almost invariably so carved, by sinking portions only of the surfaces and leaving the arrises and principal places untouched, as to preserve the original constructive forms given by the mason (Fig. 184). The employment of the drill instead of the chisel, so common in debased Roman work, was retained as a very general practice by the Greek carvers, and very often with excellent effect. The foliage of the acanthus, although imitated from the antique, quite changed its character, becoming more geometrical and conventional in its form. That which particularly distinguishes Lombard from Byzantine art is its sculpture abounding with grotesque imagery, with illustrations of every-day life, of a fanciful mythology not yet quite extinct, and allusions, no longer symbolic but direct, to the Christian creed; the latter quality a striking evidence of the triumph of the Roman Church over all iconoclastic adversaries in Greece." What is here asserted of Lombard carving is true of that in the Romanesque buildings in Germany, Scandinavia (Fig. 182), France, and to a certain extent in Great Britain, though in our own country a large proportion of the ornamental carving consists simply of decorative patterns, such as the chevron, billet, and

FIG. 182.—DOORWAY AT TIND, NORWAY.
(END OF 12TH CENTURY.)

zig-zag; and sculpture containing figures and animals is less common.

The mouldings of Romanesque buildings are simple, and at first were few in number, but by degrees they became more conspicuous, and before the transition to Gothic they

assumed considerable importance (Fig. 183) and added not
a little to the architectural character of the buildings.

FIG. 183.—MOULDINGS OF PORTAL OF ST. JAMES'S CHURCH AT KOESFELD.

Coloured decoration, especially in mosaic, was a conspicuous feature in basilican churches, and still more so
in those of the Byzantine style; such decoration in

Romanesque churches was not infrequent, but it was more commonly painted in fresco or tempera. The glass mosaic-work to be found on the walls of Early Christian churches, both basilican and Byzantine, but less frequently Romanesque, is most interesting and beautiful: "it was," says the high authority already quoted, "employed only to represent and reproduce the forms of existing objects, such as figures, architectural forms and conventional foliage, which were generally relieved with some slight indication of shading upon a gold ground—the whole being bedded in the cement covering the walls and vaults of the basilicas and churches."

"The design of both figures and ornaments was, generally speaking, very rude, though not without an occasional rising in some of the figures to a certain sublimity, derivable principally from the great simplicity of the forms and draperies and the earnest grandiose expression depicted on their countenances. The pieces of glass employed in the formation of this work are very irregular in shapes and sizes, of all colours and tones of colour, and the ground tint almost invariably prevailing is gold. The manner of execution is always large and coarse, and rarely approaches in neatness of joint and regularity of bedding to the (ancient Roman) 'opus majus vermiculatus;' yet, notwithstanding these blemishes, the effect of gorgeous, luxurious, and at the same time solemn decoration produced is unattainable by any other means as yet employed as structural embellishment. How noble and truly ecclesiastical in character are the gold-clad interiors of Monreale Cathedral, of the Capella Palatina at Palermo, of St. Mark at Venice, San Miniato at Florence, or Santi Apollinare and Vitale at Ravenna, the concurrent testimony of all travellers attests."

A finer kind of glass mosaic arranged in geometrical patterns was made use of to enrich the ambos, screens, episcopal chairs, sepulchral ornaments, and other similar fittings of churches, and was often of great beauty. A third sort of mosaic—the Alexandrine work (opus Alexandrinum)—used for pavements, has been already alluded to; this was extremely effective, but its use appears to have been less general than that of the glass mosaics for the walls.

The Architectural Character of the basilican churches may be briefly characterised as venerable and dignified, but yet cheerful and bright rather than forbidding; they are, as interiors, impressive but not oppressive, solemn but not gloomy. Comparatively little attention was paid to external effect, and there is not often much in them to strike the passer-by. The character of Byzantine interiors is far more rich, and even splendid; but it is more gloomy, and often is solemn and grand to the last degree. In many cases these churches possess fine exteriors; and for the level sky-line produced by the long straight roofs of the basilica, a more or less pyramidal composition, showing curved outlines rather than straight ones, is substituted. The architectural character of the Romanesque buildings varies extremely with the districts in which they are erected; but, generally speaking, it may be described as picturesque, and even sometimes romantic; the appearance of towers, prominent transepts, and many smaller decorative features serves to render the exteriors telling and varied, though often somewhat rude and primitive. A solid and somewhat heavy character distinguishes the interiors of some varieties of Romanesque buildings—such, for example, as our own Early Norman; but in our fully-developed and late Norman, and still more in the latest

German Romanesque churches, this disappears almost
entirely, and much beauty and even lightness of effect
is obtained, without any loss of that richness which is
characteristic of more ancient examples.

FIG. 184.—BYZANTINE BASKET-WORK CAPITAL FROM SAN MICHELE IN AFFRICISCO
AT RAVENNA.

FIG. 155.—ARABIAN CAPITAL. FROM THE ALHAMBRA.

CHAPTER XV.

MOHAMMEDAN ARCHITECTURE.

FEW revolutions more sudden, more signal, and more widespread are recorded in history than that which covered not only the East but part of the West with the Mohammedan religion and dominion. Mohammed was born either in the year 569 or 570 of the Christian era, and died A.D. 652. The year of the Hegira, the era from which Mohammedans compute their chronology, is A.D. 622, and within little more than a century from this era the Prophet was acknowledged, and the suzerainty of the Caliph recognised eastwards, in Arabia, Syria, Palestine, Egypt, and Persia, and in India as far as to the Ganges; and westwards along the north coast of Africa, in Sicily, and in Spain. It was only to be expected that such a wonderful tide of conquest and such a widespread change

of religion should before long leave its impress on the
architecture of the continents thus revolutionised; and
accordingly a Mohammedan style soon rose. This style
did not displace or override the indigenous art of the
various countries where it prevailed, as Roman architecture did in the age of universal dominion under the
Empire; it assimilated the peculiarities of each country,
and so transmuted them, that although wherever the
religion of Mohammed prevails the architecture will at
a glance confess the fact, still the local or national
peculiarities of each country remain prominent.

The Arabs, a nomadic race who lived in tents, do not
seem to have been great builders even in their cities. We
have no authentic accounts or existing remains of very
early buildings even in Mecca or Medina, as the oldest
mosques in those cities have been completely rebuilt. It
is to Egypt and Syria that we must turn for the most
ancient remaining examples of Saracenic architecture.
These consist of mosques and tombs.

Egypt.

A mosque—or Mohammedan place of worship—has
two forms. The earlier mosques are all of them of a
type the arrangement of which is simplicity itself. A
large open courtyard, resembling the garth of a cloister,
with a fountain in it, is surrounded cloister-wise by
arcades supporting timber roofs. On the side nearest
Mecca the arcades are increased to several rows in depth,
so as to cover a considerable space. This is the part in
which the congregation chiefly assembles; here a niche or
recess (termed Kibla), more or less enriched, is formed
in which the Koran is to be kept, and hard by a pulpit

is erected. For many centuries past, though not, it is believed, from the very earliest times, a minaret or high tower, from the top of which the call to prayer is given, has also been an indispensable adjunct to a mosque.

The second sort of mosque is a domed, and sometimes vaulted building of a form chiefly suggested by the Byzantine domed churches, with a central space and four short arms. This sort of mosque became almost universal in Turkey and Egypt after the capture of Constantinople by the Turks, and the appropriation to Moslem worship of Santa Sophia itself. The tombs are ornate and monumental buildings, or sanctuaries, of the same general character as the domed mosques, and often attached to them.

FIG. 186.—HORSE-SHOE ARCH.

From very early times the arches, in the arcades which have been described as virtually constituting the whole structure of the simpler sort of mosque, were pointed. Lübke claims as the earliest known and dated example of the pointed arch in a Saracenic building, the Nilometer, a small structure on an island near Cairo, which contains pointed arches that must have been built either at the date of its original construction in A.D. 719, or at latest, when it was restored A.D. 821. The Mosque of Amrou, however, which was founded very soon after the conquest of Egypt in A.D. 643, and is largely made up of

FIG. 187.—EXTERIOR OF SANTA SOPHIA, CONSTANTINOPLE. SHOWING THE MINARETS ADDED AFTER ITS CONVERSION INTO A MOSQUE.

materials obtained from older buildings, exhibits pointed arches, not only in the arcades, which probably have been rebuilt since they were originally formed, but in the outer walls, which are likely, in part at least, to be original.

Whatever uncertainty may rest upon these very remote specimens of pointed architecture, there is little if any about the Mosque of Ibn Tulun, also at Cairo, and built A.D. 885, or, according to another authority, A.D. 879. Here arcades of bold pointed arches spring from piers, and the effect of the whole structure is noble and full of character. From that time the pointed arch was constantly used in Saracenic buildings along with the semicircular and the horse-shoe arch (Fig. 186).

From the ninth century, then, the pointed arch was in constant use. It prevailed in Palestine as well as in the adjacent countries for two centuries before it reached the West, and there can be no doubt that it was there seen by the Western Crusaders, and a knowledge of its use and an appreciation of its beauty and convenience were brought back to Western Europe by the returning ecclesiastics and others at the end of the First Crusade.*

In the eleventh century the splendid Tombs of the Caliphs at Cairo were erected,—buildings crowned with domes of a graceful pointed form, and remarkable for the external decoration which usually covers the whole surface of those domes. By this time also, if not earlier, the minaret had become universal. This is a lofty tower of slender proportions, passing from a square base below to a circular form above (Fig. 187). A minaret is often divided into several stages. Each stage is then marked by a balcony, and is, generally speaking, a

---

* The First Crusade lasted from A.D. 1095 to A.D. 1099.

FIG. 188.—ALHAMBRA. HALL OF THE ABENCERRAGES.

polygon of a greater number of sides than the stage below it.

In the interiors of Saracenic buildings what is generally known as honeycomb corbelling is constantly employed to fill up corners and effect a change of plan from a square below to a circle or octagon above. This ornament is formed by the use of a series of small brackets, each course of them overhanging those below, and produces an effect some idea of which may be gathered from our illustration (Fig. 188) of the Hall of the Abencerrages in the Alhambra. The interiors when not domed are often covered by wooden or plaster ceilings, more or less richly decorated, such as are shown in the view of one of the arcades of the Mosque "El Moyed," Cairo (Fig. 189), where the horse-shoe and pointed arches can both be seen. This illustration also shows timber ties, at the feet of the arches, such as were commonly used by the earlier Saracenic builders.

The surfaces of the interiors of most Mohammedan buildings in all countries are covered with the most exquisite decorations in colour. Imitations of natural objects being forbidden by the Koran (a prohibiton occasionally, but very rarely, infringed), the Saracenic artists, whose instincts as decorators seem to have been unrivalled, fell back upon geometrical and flowing patterns and inscriptions, and upon the use of tiles (Fig. 190), mosaics, inlays, patterns impressed on plaster, and every possible device for harmoniously enriching the surfaces with which they had to deal. Several of our illustrations give indications of the presence of these unrivalled decorations in the buildings which they represent (Fig. 195). Windows are commonly filled by tracery executed in stone or in plaster, and glazed with stained glass; and

FIG. 189.—MOSQUE 'EL MOYED' AT CAIRO.

many of the open spaces in buildings are occupied by grilles, executed in wood, and most effective and rich in design.

FIG. 150.—ARABIAN WALL DECORATION.

Syria and Palestine.

Syria was one of the countries earliest overrun by the Arab propaganda, and Jerusalem was taken by the Caliph Omar as early as A.D. 637. He there built a small mosque, though not the one which commonly goes by his name. Two mosques of great antiquity and importance, but the origin of which is a matter of dispute among authorities, stand in the Haram enclosure at Jerusalem. One of these is the octagonal building called the Sakhra (Figs. 191-2), known in the Moslem world as the Dome of the Rock, popularly called the Mosque of Omar, and occupying, as is all but universally admitted, part of the site of the

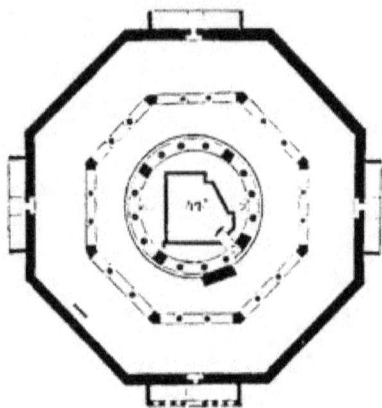

FIG. 191.—PLAN OF THE SAKHRA MOSQUE AT JERUSALEM.

Temple itself. Whether this is a "nearly unaltered Christian building of the fourth century," or a construction of Abd-el-Malek, the second Caliph, erected in the year 688, has been debated keenly; but what is beyond debate is that this structure is very Byzantine, or, to speak with more exactness, very like some of the buildings of Justinian in plan and section, and that from early times it was in the possession of the Saracens, and was regarded by them as the next most venerable and sacred spot in the world after Mecca. Much the same difference of opinion prevails as to the origin of the neighbouring mosque, El Aksah, which bears an undoubted general resemblance to an ancient basilica,

FIG. 192.—SECTION OF THE SAKHRA MOSQUE AT JERUSALEM.

though having no fewer than seven parallel avenues. This building has with equal confidence been attributed to the fourth and the seventh century. It is fortunately quite unnecessary here to do more than point out that these mosques, whatever their origin, were in use at least as early as the eighth century, and that the beautiful Dome of the Rock must have exercised a great influence on Mohammedan art, and, notwithstanding some differences of plan, may be fairly regarded as the prototype of many of the domed mosques and tombs to which allusion has been made. The decorations shown in our illustrations of the Sakhra are, it is right to observe, most of them of a date centuries later than the time of the original construction of the building.

Sicily and Spain.

The spread of Mohammedan architecture westward next claims our notice; but want of space will only permit us to mention a small though interesting group of Saracenic buildings which still remains in Sicily; the numerous specimens of the style which exist on the north coast of Africa; and the works erected by the Saracens during their long rule in Spain. The most celebrated Spanish example is the fortress and palace of the Alhambra, begun in 1248, and finished in 1314. This building (Fig. 188) has been measured, drawn, and fully illustrated in an elaborate monograph by our countryman Owen Jones, and has become popularly known by the beautiful reproduction of portions of it which he executed at the Crystal Palace, and of which he wrote an admirable description in his ' Guide-book to the Alhambra Court.' The Mohammedan architecture of Spain is here to be seen at its best; most of its features are those of Arab art, but with a distinguishing character (Fig. 193).

FIG. 193.—DOORWAY IN THE ALHAMBRA.

Two other well-known examples are, the Giralda* at

* 'Gothic and Renaissance Architecture,' p. 141.

Seville, and the Mosque at Cordova. The Giralda is a square tower, in fact a minaret on a magnificent scale, divided into panels and richly decorated, and shows a masculine though beautiful treatment wholly different from that of the minarets in Cairo. The well-known Mosque at Cordova is of the simplest sort of plan, but of very great extent, and contains no less than nineteen parallel avenues separated from one another by arcades at two heights springing from 850 columns. The Kibla in this mosque is a picturesque domed structure higher than the rest of the building. The columns employed throughout are antique ones from other buildings, but the whole effect of the structure, which abounds with curiously cusped arches and coloured decoration, is described as most picturesque and fantastic.

Persia and India.

Turning eastwards, we find in Turkey, as has been said, a close adherence to the forms of Byzantine architecture. In Persia, where the people are now fire-worshippers, the Mohammedan buildings are mostly ruined, and probably many have disappeared, but enough remains to show that mosques and palaces of great grandeur were built. Lofty doorways are a somewhat distinctive feature of Persian buildings of this style; and the use of coloured tiles of singular beauty for linings to the walls, in the heads of these great portals, and in other situations to which such decoration is appropriate, is very common: these decorations afford opportunity for the Persian instinct for colour, probably the truest in the whole world, to make itself seen.

In India the wealth of material is such that an almost unlimited series of fine buildings could be brought forward, were space and illustrations available. A large part of that vast country became Mohammedan, and in the

buildings erected for mosques and tombs a complete blend-
ing of the decorative forms in use among Hindu and Jaina
sculptors with the main lines of Mohammedan art is
generally to be found. The great open quadrangle, the
pointed arch, the dome, the minaret, all appear, but they
are all made out of Indian materials. Perhaps not the
least noteworthy feature of mosques and tombs in India
is the introduction of perforated slabs of marble in the
place of the bar-tracery which filled the heads of openings
in Cairo or Damascus. These are works of the greatest
and most refined beauty : sometimes panels of thin marble,
each pierced with a different pattern, are fitted into a
framework prepared for their reception ; at others we meet
with window-heads where upon a background of twining
stems and leaves there grow up palms or banian-trees,
their lithe branches and leaves wreathed into lines of
admirable grace, and every part standing out, owing to
the fine piercings of the marble, as distinctly as a tree
of Jesse on a painted window in a Gothic cathedral.

The dome at Bijapur, a tomb larger than the Pantheon
at Rome, and the Kutub at Delhi, a tower not unfit to
be compared with Giotto's campanile at Florence, are conspicuous among this series of monuments, and at Delhi
one of the grandest mosques in India (Fig. 194) is
also to be found. The series of mosques and tombs at
Ahmedabad, however, form the most beautiful group of
buildings in India, and are the only ones of which a
complete series of illustrations has been published.
These mosques are remarkable for the great skill with
which they are roofed and lighted. This is done by
means of a series of domes raised on columns sufficiently
above the general level of the stone ceilings, which cover
the intervening spaces, to admit light under the line
of their springing. The beauty of the marble tracery

and surface decoration is very great. Pointed arches
occur here almost invariably, and in most cases the
outline of the opening is very slightly turned upwards
at the apex so as to give a slight increase of emphasis to
the summit of the arch. The buildings are not as a
rule lofty; and though plain walls and piers occur and
contrast well with the arched features, pains have been
taken to avoid anything like massive or heavy construction.   Great extent, skilful distribution, extreme lightness, and admirably combined groupings of the features
and masses, are among the fine qualities which lend to
Mohammedan architecture in Ahmedabad a rare charm.

The religion and the art of Islam seem destined to live
and die together. Nothing (with the one exception of the
suggestion of the pointed arch to Western Europe at the
very moment when Romanesque art was ripe for a change)
has developed itself or appears likely to grow out of
Mohammedan architecture in any part of the wide field
to which the attention of the reader has been directed;
and in this respect the art of the Mohammedan is as exclusive, as intolerant, and as infertile as his religion. The
interest which it must possess in the eyes of a Western
student will rise less from its own charms than from
the fact that it first employed the pointed arch—that
feature from which sprang the glorious series of Western
Christian styles to which we give the name of Gothic.
This arch, indeed, appears to have been discovered by
the very beginners of Mohammedan architecture, at a
time when the style was still plastic and in course of
growth, and the beauty of Saracenic art is due to no
small extent to the use of it; but in the employment
of this feature the Western architect advanced much
further than the Saracen even at his best could go.
The pointed architecture of the Middle Ages, with its

daring construction, its comprehensive design, its elaborate
mouldings, and its magnificent sculptures, is far more
highly developed and more beautiful than that of the
countries which we have been describing, though in its

FIG. 105.—ENTRANCE TO A MOORISH BAZAAR.

treatment of the walls it cannot surpass, and indeed did
not often equal, the unrivalled decoration of plane surfaces
which forms the chief glory of Mohammedan art.

INDEX.

Richard Clay & Sons, Limited, London & Bungay.

www.ingramcontent.com/pod-product-compliance
Lightning Source LLC
Chambersburg PA
CBHW021508210326
41599CB00012B/1185